SpringerBriefs in Space Development

Series Editor
Joseph N. Pelton Jr.

AF796200

More information about this series at http://www.springer.com/series/10058

Olavo de Oliveira Bittencourt Neto

Defining the Limits of Outer Space for Regulatory Purposes

INTERNATIONAL®
SPACE UNIVERSITY

Springer

Olavo de Oliveira Bittencourt Neto
Santos, Brazil

ISSN 2191-8171 ISSN 2191-818X (electronic)
SpringerBriefs in Space Development
ISBN 978-3-319-16684-1 ISBN 978-3-319-16685-8 (eBook)
DOI 10.1007/978-3-319-16685-8

Library of Congress Control Number: 2015937745

Springer Cham Heidelberg New York Dordrecht London
© Springer International Publishing Switzerland 2015
This work is subject to copyright. All rights are reserved by the Publisher, whether the whole or part of the material is concerned, specifically the rights of translation, reprinting, reuse of illustrations, recitation, broadcasting, reproduction on microfilms or in any other physical way, and transmission or information storage and retrieval, electronic adaptation, computer software, or by similar or dissimilar methodology now known or hereafter developed.
The use of general descriptive names, registered names, trademarks, service marks, etc. in this publication does not imply, even in the absence of a specific statement, that such names are exempt from the relevant protective laws and regulations and therefore free for general use.
The publisher, the authors and the editors are safe to assume that the advice and information in this book are believed to be true and accurate at the date of publication. Neither the publisher nor the authors or the editors give a warranty, express or implied, with respect to the material contained herein or for any errors or omissions that may have been made.

Printed on acid-free paper

Springer International Publishing AG Switzerland is part of Springer Science+Business Media (www.springer.com)

"Two things fill the mind with ever new and increasing admiration and awe, the oftener and more steadily we reflect on them: the starry heavens above and the moral law within".

Immanuel Kant (1724–1804),
The Critique of Practical Reason[1]

[1] Immanuel Kant. *The Critique of Practical Reason*. Radford, USA: A & D, 2008. p. 111.

To Ana Carolina, my dear wife, to Lucas, our precious son, to Vera and Olavo, my adored parents, and to Tiago and Lúcio, my beloved brothers, who supported me in this new and challenging endeavor.

This Springer book is published in collaboration with the International Space University (ISU). At its central campus in Strasbourg, France, and at various locations around the world, the ISU provides graduate-level training to the future leaders of the global space community. The university offers a 2-month Space Studies Program, a 5-week Southern Hemisphere Program, a 1-year Executive MBA, and a 1-year Master's Program related to space science, space engineering, systems engineering, space policy and law, business and management, and space and society.

These programs give international graduate students and young space professionals the opportunity to learn while solving complex problems in an intercultural environment. Since its founding in 1987, the International Space University has graduated more than 3,000 students from 100 countries, creating an international network of professionals and leaders. ISU faculty and lecturers from around the world have published hundreds of books and articles on space exploration, applications, science, and development.

Contents

Chapter 1
Introduction

> Icarus, I recommend thee to keep the middle track; Lest, if thou shouldst go too low, the water should clog thy wings; if too high, the fire of the sun should scorch them. Fly between both. [Ovid (43 B.C.-17 ou 18 A.D.)] Metamorphoses[1]

Greek mythology presents as the first aviator the craftsman and architect Daedalus.[2] Confined by King Midas of Crete into the labyrinth that he once had built to imprison the Minotaur, Daedalus imagined an original way to escape from his own prison. Along with his son Icarus, he built artificial wings with feathers, attached to leather strains tied around their bodies.[3] Using these devices, legend says that Daedalus and Icarus were able to reach the skies. However, the Greek genius, acknowledging the dangers of such a novel enterprise, had warned his impetuous son that he should not fly too low, since the waves could wet his wings, nor too high, where the heat from the sun would melt the wax that held the feathers altogether.[4]

Just after they escaped from the island of Crete, Icarus disobeyed his father and gained altitude. The sky was the realm of Gods, and the young man, seduced by the brightness of the Sun, and certain of the capabilities of Daedalus' wings, decided to go beyond the limits reserved for the mortals. The tragedy anticipated by his father occurred: the feathers from Icarus' wings began to leave his arms and a terrible fall

---

[1] Ovid. *Metamorphoses*. Lawrance, USA: Digireads, 2009. p. 155.

[2] Menelaos Stephanides. *Prometeu, os Homens e Outros Mitos*. São Paulo, Brazil: Odysseus, 2008. p. 152/165.

[3] "Dédalo imediatamente pôs mão à obra. Com grande arte e habilidade, ligava as penas com cera, para fixar cada uma no lugar certo. Era um trabalho delicado, que exigia tempo e paciência, mas em poucos dias as asas já haviam tomado forma. Eram quatro, em tudo semelhantes às dos pássaros, porém muito maiores e tão bem feitas que até os deuses as teriam invejado. (...) Usando correias de couro, Dédalo prendeu um par de asas nos braços e nos ombros de Ícaro, e depois colocou as suas." Menelaos Stephanides. *Prometeu, os Homens e Outros Mitos*. São Paulo, Brazil: Odysseus, 2008. p. 159/160.

[4] William E. Burrows. *This New Ocean – The Story of the First Space Age*. New York, USA: Modern Library, 1999. p. 5.

© Springer International Publishing Switzerland 2015

O.O. Bittencourt Neto, *Defining the Limits of Outer Space for Regulatory Purposes*, SpringerBriefs in Space Development,
DOI 10.1007/978-3-319-16685-8_1

led him to his death. Daedalus, shocked by the fate of his son, managed to recover the body from the water and to bring it back to land, burying it in the island that received the name of Icaria, bathed by the Icarium Sea. When Daedalus finally reached his destination, in Sicily, he destroyed his wings. The greatest creation of the master had allowed him to get closer to the Gods, but had cost the life of his beloved son.[5]

For millennia, the dream of flying seemed impossible to be achieved.[6] However, after a spectacular string of notable triumphs that required imagination and bravery, persistence and boldness, it was possible for mankind, during the first years of the twentieth century, to conquer the skies for scientific, military and economic uses. Just a few decades later, the gravitational ties that held us to our Earth were finally mastered by rockets launched by the USSR and the USA, superpowers involved in an ideological and geopolitical conflict that recognized the strategic importance of reaching the outer space. And so, due to questionable reasons, the Space Era began.[7]

The limit that Icarus dared to defy, where aerodynamic flight became no more possible, remains undefined. Although real, it is dynamic and subjected to the forces of nature. But the advances of technology that allow mankind to repeat the flight of Daedalus, and even to boldly go as far as to the celestial bodies, lead to relevant legal discussions, especially in International Law. More conspicuously, they bring up considerations about the vertical limits of State sovereignty, the topic of this study.

What is proposed is the delimitation, through treaty rule, of the vertical frontier, tempered by passage rights for the launching and reentry of space objects, under regulation that addresses the interests of both the launching State and the territorial State that may be crossed by the flight path.

In order to fully comprehend the importance of such a proposal, it must be reckoned that the national territory of each State encompasses three dimensions. In

---

[5] "Quando Dédalo, que pode ser visto como o patrono dos técnicos de grande parte da Grécia antiga, colocou em seu filho, Ícaro, as asas que ele tinha feito, de modo que este pudesse voar e escapar do labirinto de Creta que o próprio Dédalo tinha inventado, ele disse: 'Voe moderadamente. Não voe muito alto, senão o sol derreterá a cera de suas asas e você cairá. Não voe muito baixo, senão as ondas do mar o apanharão'. O próprio Dédalo voou moderadamente, mas viu o filho, em êxtase, voando muito alto. A cera derreteu e o rapaz caiu no mar. Por alguma razão, fala-se mais de Ícaro que de Dédalo, como se as asas, em si, fossem responsáveis pela queda do jovem astronauta. Mas nada se diz contra a indústria e a ciência. O pobre Ícaro despencou nas águas; mas Dédalo, que voou moderadamente, conseguiu atingir a outra margem." Joseph Campbell. *O Poder do Mito.* São Paulo, Brazil, Brazil: Palas Athena, 1990. p. 140.

[6] "Desde o seu aparecimento na Terra, o homem passou a olhar o firmamento azul e, na sua contemplação, via as estrelas incontáveis brilharem. Naqueles êxtases, começou a sentir em su'alma o princípio da inveja dos pássaros de penas multicores a esvoaçarem, livres e felizes, pelas alturas sem fim. Possuir asas, elevar-se aos céus, penetrar no mistério do imenso infinito, descobrir a origem das coisas… eis o fruto proibido, o mais recôndito desejo do homem. Voar! Ter asas como os pássaros! E quanto mais irreal parecia o seu sonho, mais ardente era o seu desejo de torná-lo realidade num dia que não estaria longe, num futuro remoto e imprevisível". Luis Ivani de Amorim Araújo. *Curso de Direito Internacional Público.* 6. ed. Rio de Janeiro, Brazil: Forense, 1988. p. 181.

[7] José Monserrat Filho. *Direito e Política na Era Espacial.* Rio de Janeiro, Brazil: Vieira & Lent, 2007. p. 19/27.

relation to air space, that, unlike the sea, is part of the territory of every nation, its demarcation is of utmost relevance to avoid international disputes, since above a certain altitude, until now undefined, it gives way to outer space, which is free from national sovereignty and shall only be explored and used for the benefit of Mankind.

In accordance with the Chicago Convention of 1944,[8] States have exclusive and absolute sovereignty over the air space above their territories.[9] On the other hand, the Space Treaty of 1967[10] establishes that the outer space cannot be subject to national appropriation of any kind.[11] Despite configuring distinctive legal regimes that succeed each other vertically, the frontier that distinguishes air space from outer space has not been internationally delimited, even after more than 40 years of diplomatic discussions.

At the United Nations Committee for the Peaceful Uses of Outer Space (UNCOPUOS), two approaches can be identified, in relation to this issue: the first, encompassing the group of countries that became known as "spatialists", defend the demarcation of the frontier between air space and outer space based on scientific or conventional criteria; the other, of the so-called "functionalists", understand that such an endeavor is either unnecessary or impossible, suggesting that the legality of aeronautical or space activities should be accessed based on their own objectives, which shall determine the applicable legal regime.

The deadlock created by these conflicting schools of thought contributed to the development of a contradictory situation: outer space represents the vertical limit of national territories which, even though finite, extend themselves above the surface of the Earth up to a still undefined altitude.

The modern State, as it currently exists in the international order, is based upon certain elements considered essential for "statehood". Despite the divergence of Public Law scholars, it can be reasonably affirmed that these components include: (1) a people, that is, a group of individuals that are united by a legal and stable order; (2) a specific area covered by an autonomous command, subjected to a particular set of rules, home of an exclusive national jurisdiction; and (3) sovereignty, understood as the permanent and superior authority over people located in a certain territory, that allows the development of independent international relations.

In an international community composed of several States, each one as sovereign and autonomous as the other and therefore legally equal, peaceful coexistence depends upon the identification of the geographical limits of each national authority. As a consequence, political orders divided the surface of the planet into different countries, which represent the borders of their respective legal systems. This external sovereignty is reflected by the International Law principle that orders respect to the territorial integrity and political independence of States, and prohibits

[8] "Convention on International Civil Aviation".

[9] "The contracting States recognize that every State has complete and exclusive sovereignty over the airspace above its territory."

[10] "Treaty on Principles Governing the Activities of States in the Exploration and Use of Outer Space, Including the Moon and Other Celestial Bodies."

[11] "Outer space, including the Moon and other celestial bodies, is not subject to national appropriation by claim of sovereignty, by means of use or occupation, or by any other means."

interference on matters essentially within the domestic jurisdiction of a specific nation. As stated by Michel Foucher, there is no national identity without clearly provided and respected frontiers.[12]

Above the surface of the Earth, national territories meet a true limit to sovereignty, provided by Space Law. At a certain altitude, still to be provided, two legal systems enter in a constant and necessary contact. The air space/outer space frontier has a true but undetermined border.

In order to support a proposal for the resolution of one of the oldest questions in Space Law, this text presents multidisciplinary research in order to comprehend the conflicting opinions about the vertical limits of State sovereignty. Lessons on Comparative Law shall be reviewed, in order to appreciate the international legal regimes applicable to different territories. Developed in order to encompass rules referring to a certain location, through *ratione loci* competences, the Space Law, the Air Law and the Law of the Seas have an inherent and necessary geographical scope. Nevertheless, all of these legal systems have necessarily to deal with the provision of limits applicable to regions that, at least for the current stage of human knowledge, are virtually indistinguishable.

The present study represents a brief, but reviewed, account of my doctorate thesis presented before the University of São Paulo, in 2011, which granted me a PhD title *summa cum laude*. It was concluded after a period of research conducted at Leiden University, made possible by a Coimbra Group scholarship. Herein are included relevant elements for further academic study on this subject matter. It is expected that the proposal hereby introduced may, in a way or another, contribute to provide legal justification for a compromised solution before the UNCOPUOS, forum that has, for so long, devoted singular attention to the definition and delimitation of outer space.

This book could not be concluded without the help of many people. In particular, I owe a great debt of gratitude to Professor José Monserrat Filho, from the Brazilian Space Agency, my true "mentor" in Space Law, who taught me everything I know in this field; to Professor Paulo Borba Casella, from the University of São Paulo, who supervised both my Masters dissertation and my PhD thesis with utmost care and attention, sharing his knowledge and wisdom; to Professor Tanja-Masson Zwaan, from Leiden University, who magnificently supervised this project, making me believe I could offer a relevant contribution to a complex legal problem; and finally, to Professor Marianne Wiesebron, from Leiden University, who supported my researches in the Netherlands, while providing everlasting academic advises. I consider this research a true "teamwork"; to all of those who accompanied me throughout this journey, my deepest thanks.

[12] "As fronteiras são descontinuidades territoriais, com a função de marcação política. Nesse sentido, trata-se de instituições estabelecidas por decisões políticas, projetadas ou impostas, e administradas por textos jurídicos: as leis de um Estado soberano em seu interior, o Direito Internacional Público como lei comum da coexistência dos Estados, mesmo quando estes se desfazem, porque os tratados territoriais são os únicos pelos quais a sucessão de Estado é automática (…) não há identidade sem fronteiras. A ordem política moderna implica o reconhecimento, pelos outros, de fronteiras de Estado demarcadas, com base territorial e soberana". Michel Foucher. *Obsessão por Fronteiras*. São Paulo, Brazil: Radical Livros, 2009. p. 22.

Chapter 2
Air Space and Outer Space

In the beginning, it was necessary not only to fight against the elements, but also the prejudices; the direction of balloons and, later, the mechanical flight, were considered 'unsolvable problems' [Alberto Santos-Dumont (1873–1932)], O Que Eu Vi, o Que Nós Veremos[1]

The differences between the legal regimes applicable to air space and outer space are of a fundamental order: while Air Law is based on considerations of sovereignty, Space Law overtly forbids any form of national appropriation.[2] Widely accepted treaties provided those rules, which can even be recognized as of a fundamental character to Air Law and Space Law. Therefore, the different standards contributed to the creation of two immiscible legal systems, which arguably succeed each other above the surface of the Earth, at a still to be determined altitude.

Political, economic and military considerations influenced the legislative paths that were to be followed by Air Law and Space Law. When the use of aircraft as weapons of war was consolidated during the First World War (1914–1918), against the wishes of such pioneers of aviation as the Brazilian Alberto Santos-Dumont (1873–1932), the voices that defended the "freedom of the air" were overpowered by a pragmatic claim in favor of vertical sovereignty. Airplanes were, from then on, widely recognized as strategic tools for the observation of enemy troops, for defense against foreign threats and as important attacking weapons. At the same time, due to the development of aeronautic technology, as the exploration of commercial air routes for transportation became feasible, economic considerations supported the protection of national air companies. Consequently, aviation would be submitted to a legal regime that respected the sovereignty of the territorial State.

[1] "A princípio, tinha-se que lutar não apenas contra os elementos mas também contra os preconceitos: a direção dos balões e, mais tarde, o voo mecânico, eram problemas 'insolúveis'." Alberto Santos-Dumont. *O Que Eu Vi, o Que Nós Veremos*. São Paulo, Brazil: Hedra, 2000. p. 17.

[2] I. H. Diederiks-Verschoor. "Differences Between Air Law and Space Law". Recueil des Cours, 172, (1981-III), p. 320.

© Springer International Publishing Switzerland 2015

O.O. Bittencourt Neto, *Defining the Limits of Outer Space for Regulatory Purposes*, SpringerBriefs in Space Development,
DOI 10.1007/978-3-319-16685-8_2

The strategic importance of outer space also had a definitive effect over its own set of international rules. There was no interest by the superpowers that dueled during the Cold War (1945–1989),[3] the USA and USSR, to prohibit the use of intercontinental ballistic missiles, important weapons in case of a foreseeable nuclear war, which, during their parabolic trajectory to the final target, pass above the territory of several States at altitudes well beyond the most significant atmospheric layers.[4] Likewise, the use of observation satellites was desired by both space powers, in a kind of "authorized espionage" that would be fundamental for the preservation of the *status quo* and disarmament proposals that became the rule during the period of deterrence, by the end of the 1960s. Those are some of the major reasons why proposals favorable to the vertical extension of national sovereignty without limits did not prosper among delegates attending the international meetings to prepare treaties regarding this new field of International Law.

It is important to reckon, as done by Ruwantissa I. R. Abeyratne, that those differences on sovereignty between Air Law and Space Law that eventually characterize one as the antithesis of the other have produced important legal consequences: "while the implementation of Air Law is heavily influenced by municipal law, space law is solely grounded on legal principles binding on the community of nations. Principles of Public International Law therefore play an exclusive part in the application of space law principles."[5]

Air Space: Complete and Exclusive Sovereignty

The territorial integrity principle, necessary for the legal concept of sovereignty, is not restricted to its horizontal dimension; in fact, its vertical application received special attention from International Law ever since aviation became feasible, by the beginning of the twentieth century. Due to constant technological evolution, States progressively established a far reaching regulation, based not only in military, but also commercial considerations, providing a rather complex international system of rules. Due to that, the legal *corpus* of Air Law can nowadays be considered one of

[3] "Following World War II, the foreign policies of the United States and the Soviet Union interacted with the chaotic and fluid state of international relations to produce the Cold War. Understanding the impact of World War II on the international system and its members is crucial to understanding the origins of the Cold War. World War II accelerated fundamental changes in the global distribution of power, in weapons technology, in the balance of political forces among and within nations, in the international economy, and in relations between the industrial nations and the Third World. In addition, the diplomatic and military decisions made during the war had a profound impact on the shape of the postwar world." David S. Painter. *The Cold War: an International History*. New York, USA: Routledge, 1999. p. 4.

[4] "Um foguete que servisse de propulsor para uma nave espacial também poderia fazer o mesmo com um míssil carregando uma ogiva nuclear." Geoffrey Blainey. *Uma Breve História do Século XX*. São Paulo, Brazil: Fundamento Educacional, 2008. p. 211.

[5] Ruwantissa I. R. Abeyratne. *Frontiers of Aerospace Law*. Burlington, USA: Ashgate, 2002. p. 5.

the more comprehensive ones, as far as uses of a territory are concerned, encompassing treaties vastly welcomed by the international community, as well as further bilateral agreements between States.[6]

The air space has particular features that represented relevant challenges to International Law. In accordance with Albino de Azevedo Soares:

> Although aviation is much more recent than maritime navigation, the questions that it presents, however, are not simpler. In fact, hardly aeronautical communications between two countries can be performed without the over flight of different States; if not all States have maritime dominium, there is not a single one without air space; in the later, frontiers are not demarked in a visible way; aeronautical safety demands uniformity of traffic signals.[7]

Legal considerations regarding the use of air space precede the development of aircraft. The overflight of balloons beyond national borders brought the attention of publicists still in the nineteenth century, as indicated by the Brussels Declaration of 1874 about Law and Customs of War, which provided that "persons sent in balloons for the purpose of carrying dispatches and, generally, of maintaining communications between the different parts of an army or a territory" should not be considered spies (article 22).[8] During the First Peace Conference held at the Hague in 1899, the use of balloons as weapons of war was suspended for a specific period by the delegation, due, among other reasons, as observed by David Johnson, to the remaining doubt about its strategic capability for attacks, so that the military use of balloons was reserved to observation outposts.[9]

International legal considerations gained momentum after the first aviators dominated the technology of heavier than air machines, capable of sustaining flight by a fine combination of mechanical propulsion and aerodynamic lift. Concrete problems in relation to security and transportation demanded cooperation among States for their proper solution.[10]

---

[6] "The very concept of international air law may give rise to controversy, since its content entails a vastly more complex system of law than any other system of law. (…) Thus the law needs to build up a great apparatus of manageable concepts in order to provide a workable framework for international air law." Chia-Jui Cheng. "New Sources of International Air Law", *In*: Chia-Jui Cheng (ed.). *The Use of Air Space and Outer Space for All Mankind in the 21st Century.* The Hague, The Netherlands: Kluwer, 1995. p. 277.

[7] "Sendo a aviação muito mais recente do que a navegação marítima, não são, todavia, as questões que ela suscita menores do que as desta. De fato, dificilmente as comunicações aéreas entre países distantes se podem fazer sem o sobrevoo de Estados diferentes; se nem todos os Estados possuem domínio marítimo, não há nenhum sem espaço aéreo; neste, as fronteiras não são demarcadas duma forma visível; a segurança da navegação exige uniformidade de sinalização." Albino de Azevedo Soares. *Lições de Direito Internacional Público.* 4. ed. Coimbra, Portugal: Coimbra Editora Limitada, 1988. p. 259.

[8] David Johnson. *Rights in Air Space.* Manchester, England: Manchester University Press, 1965. p. 10.

[9] "At the Hague in 1899 it was felt that the use of balloons for the purpose of discharging projectiles was attended by too many uncertainties to be permitted." David Johnson. *Rights in Air Space.* Manchester, England: Manchester University Press, 1965. p. 11.

[10] Michael Milde. *International Air Law and ICAO.* Utrecht, The Netherlands: Eleven International Publishing, 2008. p. 6/8.

A new body of rules, based in international treaties, was conceived for the "vertical territory", often times anticipating technological capability. The designation of this legal field, nevertheless, has been subjected to an argument that remains even today between those that defend the designation "Air Law" to others who suggest different alternatives, for instance, "Aeronautical Law" or "Aviation Law". Nevertheless, it must be acknowledged that the term "Air Law" is commonly used in international fora and has, in its favor, the benefit of comprehending all aspects of human activities in air space. It was used by I. H. Ph. Diedericks-Verschoor, who defined Air Law as "a body of rules governing the use of airspace and its benefits for aviation, the general public and the nations of the world."[11]

Rather interesting to notice that such term was originally applied only for legal regulation regarding air navigation, but later welcomed a much broader field of action, as indicated by Michael Milde:

> The very term air law (droit aerien, derecho aéreo, Luftrecht, vozdushnoye pravo) is controversial and imprecise but it has been used in practice for over a century. It was arguably coined, in its French form, by Professor Ernest Nys of the University of Brussels in 1902 in its report to the Institute de Droit International in the subject 'droit des aerostats'. (…) It is safe to conclude that the term 'air law' from its inception was confined only to the legal regulation of social relations generated by the aeronautical uses of the air space.[12]

In Brazil, Adherbal Meira Mattos also adopted the term "Air Law", affirming that it "comprehends the group of international norms that rule the air space and its use. Therein, are included problems regarding navigation, radiotelegraphy and radiotelephony."[13] The same reasoning can be found in the works of other important Brazilians, including Paulo Borba Casella[14] and Luis Ivani de Amorim Araujo.[15] Explained Celso D. de Albuquerque Mello that "the expression 'air law' has been criticized for being 'vague and imprecise', since it includes the air space and all its uses, like navigation, radiotelegraphy, etc. However, as far as International Law is concerned, it provides the great advantage of encompassing in a single term all the different uses of the air space, whose rules shall depend upon the regulation of air space itself."[16]

---

[11] I. H. Diedericks-Verschoor. *An Introduction to Air Law*. 6. ed. Deventer, The Netherlands: Kluwer, 1993. p. 1.

[12] Michael Milde. *International Air Law and ICAO*. Utrecht, The Netherlands: Eleven International Publishing, 2008. p. 1/2.

[13] "Direito Aéreo compreende o conjunto de normas internacionais que regulam o espaço aéreo e sua utilização. Nesta, estão incluídos problemas relativos à navegação, à radiotelegrafia e à radiotelefonia." Adherbal Meira Mattos. *Direito Internacional Público*. 2. ed. São Paulo, Brazil: Renovar, 2002. p. 211.

[14] Paulo Borba Casella. *Direito Internacional dos Espaços*. São Paulo, Brazil: Atlas, 2009. p. 497/499.

[15] Luis Ivani de Amorim Araújo. *Curso de Direito Internacional Público*. 6. ed. Rio de Janeiro, Brazil: Forense, 1988. p. 184/185.

[16] "A expressão 'direito aéreo' tem sido criticada por ser 'vaga e imprecisa', isto é, ela abrange o meio aéreo e todas as suas utilizações, como a navegação, a radiotelegrafia, etc. Entretanto, no DI [Direito Internacional] traz a grande vantagem de abranger em um único termo as diferentes utilizações do

The other term usually employed to identify this legal branch, "Aviation Law", is naturally encompassed by the broader designation "Air Law", since the former provides rules regarding aeronautic navigation, while the later addresses all uses of air space. Thus, the apparent conflict of terminology, commonly refereed in Romance language countries, in fact never takes place, as clarified by J. C. Sampaio Lacerda.[17]

Air Law "took off" among a debate those that defended freedom of air space and those that advocated national sovereignty over the column of air above States' territories. Such discussion emerged after it was clearly realized that air navigation provided relevant challenges to national security.[18] As explained by Federico N. Videla Escalada, "only when air traffic began to gain ground and to demand the attention of States, did they – on account of their own interests – understand that it was imperative to regulate duly the activities of the newly arisen means of communication which, in addition, offered very special characteristics as an offensive and defensive weapon in the event of an outbreak of hostilities."[19]

The French renowned legal scholar Paul Fauchille, in his study "Le Domaine Aéreian et Le Régime Juridique de les Aerostats" of 1901, raised the flag of freedom of air, balanced by the right of self-defense of the territorial State. To him, the thesis that should prevail was the one which claimed that "l'air est libre".[20] Fauchille, by the beginning of the twentieth century, suggested to the International Law Institute (ILA) the provision of an international code of civil aviation supporting his proposal, based on such singular opportunity of Law anticipating the technological evolution.[21] In 1906, ILA, during its meeting in Gand, Belgium, adopted a resolution

---

espaço aéreo, cujas regulamentações vão depender da própria regulamentação do espaço aéreo." Celso D. de Albuquerque Mello. Curso de Direito Internacional Público, Volume II. 15. ed., Rio de Janeiro, Brazil: Renovar, 2004. p. 1.308.

[17] "Direito aéreo e direito aeronáutico. É preciso distinguir-se o sentido em que devem ser empregadas essas duas expressões. Direito aéreo corresponde a um campo de ação mais amplo: abrange não só as normas relativas à locomoção aérea, como ainda as necessárias a regulamentação das diversas atividades utilizadas no espaço aéreo, alcançando as aplicações das invenções recentes: condutos elétricos, telégrafos, radiotelegrafia, radiotelefonia e a televisão. Direito aeronáutico será, então, o complexo de normas jurídicas relativas à navegação feita pelo ar. Seu domínio restringe-se, pois, ao fenômeno da locomoção aérea. Não, apenas, aos transportes aéreos, pois, neste caso, estariam excluídas outras atividades da navegação aérea (esportiva, por exemplo)". J. C. Sampaio de Lacerda. *Curso de Direito Comercial Marítimo e Aeronáutico (Direito Privado da Navegação)*. Rio de Janeiro, Brazil: Freitas Bastos, 1957. p. 387/388.

[18] "Temos que entender que uma das principais responsabilidades do Direito Aeronáutico é impedir a invasão de territórios de outros países, já que esse é um motivo precursor da regulamentação jurídica da aeronavegação". Paulo Henrique de Souza Freitas. *Responsabilidade Civil no Direito Aeronáutico*. São Paulo, Brazil: Juarez de Oliveira, 2003. p. 37.

[19] Federico N. Videla Escalada. *Aeronautical Law*. Alpheen aan den Rijn, the Netherlands: Sijthoff & Noordhoff, 1979. p. 54.

[20] "(...) son immensité, as fluidité, as mobilité le rend [le domanie aérien] comme la mer insusceptible de souveraineté autant que de propriété". Paul Fauchille. *Le Domaine Aéreian et Le Régime Juridique de les Aerostats*. Paris, France: Pendone, 1901. p. 19.

[21] The proposal was offered at the Neuchatel meeting, in 1900, and later succeeded by a project developed by Fauchille himself in 1902, during the Brussels meeting. Luis Ivani de Amorim Araújo. *Curso de Direito Internacional Público*. 6. ed. Rio de Janeiro, Brazil: Forense, 1988.

regarding the legal aspects of air navigation and radiotelegraphy which included, in its first article, the following statement: "The air is free. States have over it, in times of peace or war, just the rights necessary to their own protection".[22]

In a remarkable evolution of his thesis, Fauchille progressively understood that the national sovereignty should extend itself until a vertical limit, originally proposed at 300 m of altitude; beyond such standard, up to 1,500 m, would be found an intermediary zone, where States should retain the right to restrict flights, in order to assure their self-defense. Further on, the international air space would commence, following the customary legal regime applicable to the high seas.[23] Later, Fauchille changed those standards, eventually fixing national sovereignty based on the height of the Eiffel Tower, that is, 330 m.[24]

However, the fast evolution of aviation, enhanced by an international arms race that would lead to the use of aircraft as weapons of war, clearly demonstrated the strategic importance of controlling national airspace. By the end of the First World War, it was evident to everyone that no trench on the ground could effectively hinder the assault of enemy airplanes. Also, the potential of commercial aeronautic navigation began to be realized, leading States to understand the importance of protecting their national air companies and flight routes.

As indicated by Wencelas J. Wagner:

> The fantastic development of aviation during the World War I revealed to the nations the tremendous possibilities of transportation by air and proved that this way of travel was the most international of all. On the other hand, the war demonstrated the important role aviation assumed in military operations, and brought up the idea that appropriate precautions should be taken against this means of transportation which becomes a device for destruction during hostilities.[25]

In just a few years, the proposal of Fauchille fell from grace and became seen as too romantic by the international community. The constant development of aviation brought a common conviction to world leaders that their national security would be dangerously impaired if effective control of airspace was not achieved. In fact, A. Vereschaguin affirmed that States never truly considered that their national air space were opened to foreign aircraft, right from the beginning clearly stating their

---

p. 192. I. H. Diedericks-Verschoor. *An Introduction to Air Law*. 6. ed. Deventer, The Netherlands: Kluwer, 1993. p. 1.

[22] "L'air est libre. Les fetats n'ont sur lui, en temps de paix et en temps de guerre, que les droits necessaires a leur conservation."

[23] Paul Fauchille. *Le Domaine Aéreian et Le Régime Juridique de les Aerostats*. Paris, France: Pendone, 1901.

[24] "Posteriormente, isto é, em 1910, o mesmo jurista preferiu fixar tal limite em 500 metros do solo, tendo em vista o progresso alcançado pelos aparelhos fotográficos, conforme deixou esclarecido em seu 'La circulation aérienne et les droits des États en temps de paix' até que, ao escrever seu 'Traité de Droit International Public', em 1925, tomando por base a altura da Torre Eiffel – a mais alta das construções, na época – fixou tal limite em 330 metros". Luis Ivani de Amorim Araújo. *Curso de Direito Internacional Público*. 6. ed. Rio de Janeiro, Brazil: Forense, 1988. p. 197/198.

[25] Wenceslas J. Wagner. *International Air Transportation as Affected by State Sovereignty*. Brussels, Belgium: Bruylant, 1970. p. 37.

exclusive right to that part of atmosphere.[26] Added Celso D. de Albuquerque Mello that, in his opinion, the thesis espoused by Fauchille was unacceptable, since wrongly confusing the air, a true *res communis*, with the air space subjected to national sovereignty.[27,28] In the same sense, can be found the opinions of Luis Ivani de Amorim Araujo and Luis Tapia Salinas,[29] as well as Morton A. Kaplan and Nicholas de B. Katzenbac.[30]

Indeed, the First Pan-American Aeronautical Conference held in Santiago, Chile, in 1916, defended categorically that States had sovereign rights over their respective air spaces. For the same reasons, in February 8th, 1919,[31] just after the end of the First World War and with the establishment of the first regular air route connecting Paris to London, discussions about a comprehensive international treaty about Air Law[32] favored the thesis of John Westlake. This British legal scholar who advocated the sovereignty of States in relation to the air space above their territories, tempered by the right of innocent passage.[33] As observed by Paulo Borba Casella, there is a self-evident connection of that proposition with the "ancient principle, attributed to the school of glossers of the XVIIIth century, with Accursio (who died in 1263) (...): cujus est solum, ejus est usque ad coelum – and it can be included et ad inferos, understanding that belong to the owner of the land what it is above it until the

[26] "La práctica internacional en los años precedentes a la primera guerra mundial y, sobre todo, durante el propio período bélico, confirmó que los Estados no se inclinaban de modo alguno a considerar su espacio aéreo como objeto de uso general y desde un principio anunciaron inequívocamente su derecho exclusivo a esa parte de la atmosfera". A. Vereschaguin. "Derecho Aéreo Internacional", *In*: G. Tunkin (coord.). *Curso de Derecho Internacional: Manual*. Moscow, Russia: Progresso, 1979. p. 71.

[27] "O ar é realmente inapropriável, mas o espaço aéreo é apropriável." Celso D. de Albuquerque Mello. Curso de Direito Internacional Público, Volume II. 15. ed., Rio de Janeiro, Brazil: Renovar, 2004. p. 1.309.

[28] Luis Ivani de Amorim Araújo. *Curso de Direito Internacional Público*. 6. ed. Rio de Janeiro, Brazil: Forense, 1988. p. 191.

[29] Luis Tapia Salinas. *Manual de Derecho Aeronáutico*. Barcelona, Spain: Bosch Casa Editorial, 1944. p. 41.

[30] "Na verdade, em retrospecto, o que surpreende é o vigor dos argumentos apresentados por alguns partidários da liberdade do espaço. Desde o início ficou evidente a importância militar, no mínimo para fins de observação, das atividades aéreas e, em consequência, os Estados preocuparam-se em regular e controlar os processos que tratavam de voos estrangeiros sobre seus territórios. Outro argumento favorável era o perigo dos voos, não só para os que dele participavam como para os que permaneciam em terra, e geralmente se pensava, também por esta razão, que o voo deveria ser regulamentado pelo Estado subjacente." Morton A. Kaplan and Nicholas de B. Katzenbach. *Fundamentos Políticos do Direito Internacional*. Rio de Janeiro, Brazil: Zahar, 1964. p. 169.

[31] I. H. Diedericks-Verschoor. An Introduction to Air Law. 5. ed. Deventer, the Netherlands: Kluwer, 1993. p. 2.

[32] Reference shall be made to the 1910 Paris International Conference regarding international air navigation, that could not conclude an international treaty mostly due to the conflict of opinions regarding the legal nature of air space. Michael Milde. *International Air Law and ICAO*. Utrecht, the Netherlands: Eleven International Publishing, 2008. p. 9.

[33] Celso D. de Albuquerque Mello. Curso de Direito Internacional Público, Volume II. 15. ed. Rio de Janeiro, Brazil: Renovar, 2004. p. 1.309.

Heaven and below it until the 'Hell'. The maxim may have prevailed in relation to the space subjected to Air Law, but years later would be restricted by the development of Space Law, that, by its turn, does not and could not claim sovereignty over the outer space."[34]

The defense of national sovereignty over the air space was reflected in the Paris Convention of 1919, which introduced the first comprehensive international regulation of Air Law focused on its civil and commercial aspects. Signed in October 3rd, 1919 by 27 States, including Brazil, the treaty clearly established in its first article: "the High Contracting Parties recognize that every Power has complete and exclusive sovereignty over the air space above its territory."

The Paris Convention's relevance cannot be undermined, since it edified, through a binding legal regime, a whole new branch of International Law, as concluded by Michael Milde:

> The Convention is the historically first multilateral instrument of international law relating to air navigation. It helped to formulate also the principles of the domestic law of contracting States, many of whom by 1919 did not have any laws governing aviation.[35]

As a matter of fact, it can be stated that, especially due to the Paris Convention, the absolute and exclusive national sovereignty over air space has become a general principle of law. "One can indicate", affirmed Hildebrando Accioly, "that every or almost every municipal laws of different States, regarding air navigation, are based on the doctrine of sovereignty of the State over the air space above its respective territory".[36]

A brief explanation shall be provided: Fauchille's theory of freedom of the air, then abandoned, must not be mistaken with the "open-skies" policy, applicable to commercial flights. As well explained by Peter P. C. Haanappel, "an open-skies policy – a misnomer, in the public international law sense of the word – does not mean that sovereignty in national airspace is abolished; it means that a country unilaterally, for domestic air transport, or bilaterally or multilaterally, for international

---

[34] "Evidente a conexão com o antigo preceito, atribuído à escola dos glosadores do século XVIII, com ACCURSIO (falecido em 1263) (...): cujus est solum, ejus est usque ad coelum – e se pode acrescentar et ad inferos, entendendo que pertence ao proprietário do solo, o que está acima dele até o céu e abaixo dele até o 'inferno'. O preceito pode ter prevalecido, em relação ao espaço de competência do Direito Aeronáutico, mas, anos depois, seria restringido pelo desenvolvimento do Direito Espacial, que, por sua vez, não adota nem seria viável pretender a soberania do espaço territorial." Paulo Borba Casella. *Direito Internacional dos Espaços*. São Paulo, Brazil: Atlas, 2009. p. 504. Also: Malcolm N. Shaw. International Law. 3. ed. Cambridge: Cambridge University Press, 2003. p. 464. John Cobb Cooper. "Roman Law and the Maxim Cujus Est Solum in International Air Law". *McGill L. J.*, n°. 23. Montreal, Canada: 1955.

[35] Michael Milde. *International Air Law and ICAO*. Utrecht, the Netherlands: Eleven International Publishing, 2008. p. 10.

[36] "Podemos assinalar que todas ou quase todas as leis internas dos diferentes Estados, relativas à navegação aérea, são baseadas na doutrina da soberania do Estado sobre o espaço aéreo acima do respectivo território." Hildebrando Accioly. *Tratado de Direito Internacional Público*. Vol. II. 3. ed. São Paulo, Brazil: Quartier Latin, 2009. p. 330.

air transport, adopts a commercial regime based upon a maximum of free market forces and minimum of (inter) governmental intervention."[37]

Returning to the Paris Convention, another of its great achievements was the creation of an intergovernmental organization responsible for the regulation of air navigation, reserving special attention to technical procedures, in order to guarantee the safety of international flights, named CINA ("Comission Internationale de la Navigation Aérienne"), under the wings of the League of Nations.[38]

Also, one must stress that this agreement, in its article 15, provided the right of innocent passage for aircraft over the air space of other contracting parties without landing, as long as in accordance with other relevant treaty rules, including the respect of routes fixed by the overflown State. The generality of the wording used in this article led to different interpretations, notably in relation to who should have the right to determine if a passage was peaceful or not.[39]

Two regional treaties concluded during the 1920s deserve attention as well, since they reiterated the rules internationally provided by the Paris Convention. The first one, the Ibero-American Convention on Air Navigation, between Spain, Portugal and several Latin American countries, was concluded in 1926 in Madrid, indicating that "each power has complete and exclusive sovereignty over the atmospheric space above its territory."[40] Two years later, the Pan American Union, predecessor of the Organization of American States (OAS), promoted the Pan American Convention on Commercial Aviation, opened for signature in Havana, once again stating that "each State has complete and exclusive sovereignty over the air space above its territory and continental waters."[41] Innocent passage through air space during times of peace was equally considered by both treaties, though impaired by the same complications of the Paris Convention.[42]

Several air routes were created throughout the globe from the 1920s on, allowing increasingly more comprehensive international communication. Engineering advances allowed aircraft to reach higher altitudes, free from the major turbulences

---

[37] Peter P. C. Haanappel. *The Law and Policy of Air Space and Outer Space: a Comparative Approach*. The Hague, the Netherlands: Kluwer Law International, 2003. p. 4.

[38] Michael Milde. *International Air Law and ICAO*. Utrecht, the Netherlands: Eleven International Publishing, 2008. p. 11.

[39] Hildebrando Accioly. *Tratado de Direito Internacional Público*. Vol. II. 3. ed. São Paulo, Brazil: Quartier Latin, 2009. p. 333.

[40] "Cada potência tem a soberania completa e exclusiva sobre o espaço atmosférico acima de seu território."

[41] "Cada Estado tem completa e exclusiva soberania sobre o espaço aéreo acima de seu território e águas continentais."

[42] "Mas, o que vem a ser passagem inocente? (…) se há normas determinantes emanadas dos Estados sobre a possibilidade, a liberdade de aeronaves estranhas sobrevoarem territórios e neles pousarem na sua missão de transportar passageiros e mercadorias de um para outro Estado, esse voo é considerado inocente, essa travessia é inofensiva, porque dentro do estabelecido nas convenções entre Estados. Mas, cabe ao Estado sobrevoado ou utilizado como ponto de pouso, isto é, o Estado ao qual pertence o espaço aéreo esclarecer se a passagem tem ou não a característica requerida." Luis Ivani de Amorim Araújo. *Curso de Direito Internacional Público*. 6. ed. Rio de Janeiro, Brazil: Forense, 1988. p. 195/196.

close to the surface, with a greater capacity to cover distance. National air companies, both public and private, started to compete for customers, along the way providing better services and cheaper passages. Therefore, by the Second World War, aviation had already conquered a fair share of the international transportation market. The Paris Convention's rules seemed increasingly more outdated in face of such new, commercially related, challenges.

Accordingly, even before the end of the fighting in the Pacific, the American president Franklin Delano Roosevelt (1881–1945) answered a proposition of its ally, British Prime Minister Winston Churchill (1874–1965) by inviting delegations of several fellow countries to review the international regulation applicable to air space and aviation. Thus on December 7th, 1944, the Chicago Convention on International Civil Aviation was signed as the major document of said conference. This was followed by two additional agreements concerned with regular air services (regarding air traffic and air transport), and 18 other annexes encompassing technical regulation connected to the main agreement.[43]

At Chicago it was decided that a new international organization should be established to regulate civil aviation, named ICAO (International Civil Aviation Organization), with headquarters in Montreal.[44] International jurisdiction was granted to the ICAO in accordance with article 47 of the Chicago Convention treaty.[45] "The ICAO became a special body of the United Nations soon after it was created", remarked Paulo Borba Casella, "being recognized for the importance of the international legislation adopted by this organization, which finds undeniable acceptance and absolute uniformity of application, in all its works".[46] In a relevant study about the importance of ICAO to Air Law, Michael Milde

[43] I. H. Diedericks-Verschoor. An Introduction to Air Law. 5. ed. Deventer, the Netherlands: Kluwer, 1993. p. 10/11.

[44] The ICAO objectives were provided by article 44 of the Chicago Convention: "The aims and objectives of the Organization are to develop the principles and techniques of international air navigation and to foster the planning and development of international air transport so as to: (a) Insure the safe and orderly growth of international civil aviation throughout the world; (b) Encourage the arts of aircraft design and operation for peaceful purposes; (c) Encourage the development of airways, airports, and air navigation facilities for international civil aviation; (d) Meet the needs of the peoples of the world for safe, regular, efficient and economical air transport; (e) Prevent economic waste caused by unreasonable competition; (f) Insure that the rights of contracting States are fully respected and that every contracting State has a fair opportunity to operate international airlines; (g) Avoid discrimination between contracting States; (h) Promote safety of flight in international air navigation; (i) Promote generally the development of all aspects of international civil aeronautics."

[45] "The Organization shall enjoy in the territory of each contracting State such legal capacity as may be necessary for the performance of its functions. Full juridical personality shall be granted wherever compatible with the constitution and laws of the State concerned."

[46] "A OACI tornou-se órgão especializado das Nações Unidas logo após sua criação notabilizando-se pela importância da legislação internacional adotada no âmbito desta organização, que encontra aceitação inconteste e absoluta uniformidade de aplicação, em todos os seus trabalhos". Paulo Borba Casella. *Direito Internacional dos Espaços*. São Paulo, Brazil: Atlas, 2009. p. 510/511.

highlights its capacity of action in the international community, effectively universal nowadays.[47]

The Chicago treaty reinstated national sovereignty over the respective air space, as provided by its article 1,[48] with the additional indication, in article 2, that the competence is extended to land areas and territorial waters adjacent thereto.[49] So, as observed by I. H. Diedericks-Verschoor, the rule of national sovereignty continued to be the legal standard applicable to international civil aviation.[50]

Although the rules established in Chicago were more concerned with civil aircraft, article 3 item (c) affirms that governmental aircraft (including and especially the military ones) shall fly over the territory of other States only with previous authorization, otherwise being considered invaders and, therefore, subject to applicable self-defense measures.[51] The same applies to landing procedures: if no authorization is granted, governmental aircraft shall be considered in breach of local national sovereignty. Some famous cases of international disputes regarding overflight by foreign state aircraft were recalled by William R. Slomanson, including an interesting one, that deserves mention: "France, for example, legally denied US President Ronald Reagan's request that US military aircraft fly over France during a 1986 retaliatory bombing mission in Libya. US warplanes thus were required to fly a circuitous and much longer route around France's territorial airspace – through the Straits of Gibraltar – to gain access to Libya via the western entrance to the Mediterranean Sea."[52]

As acknowledged by Peter Malanczuk, "it is a serious breach of international law for a state to order its aircraft to violate the air space of another state."[53] The air espionage conducted conspicuously during the first years of the Cold War created many dangerous diplomatic incidents as a result of this, as when an American military aircraft U-2 was intercepted in action over Soviet territory in 1960, bringing the pilot, even though rescued alive, subjected to judgment for espionage; he was only

---

[47] Michael Milde. *International Air Law and ICAO*. Utrecht, the Netherlands: Eleven International Publishing, 2008. p. 124.

[48] "The contracting States recognize that every State has complete and exclusive sovereignty over the airspace above its territory."

[49] "For the purposes of this Convention the territory of a State shall be deemed to be the land areas and territorial waters adjacent thereto under the sovereignty, suzerainty, protection or mandate of such State."

[50] I. H. Diedericks-Verschoor. An Introduction to Air Law. 5. ed. Deventer, the Netherlands: Kluwer, 1993. p. 9.

[51] "Nenhuma aeronave governamental pertencente a um estado contratante poderá voar sobre o território de outro Estado, ou aterrissar no mesmo sem autorização outorgada por acordo especial ou de outro modo e de conformidade com as condições nele estipuladas."

[52] William R. Slomanson. *Fundamental Perspectives on International Law*. 4. ed. Belmont, USA: Wadsworth/Thomson Learning, 2003. p. 282.

[53] Peter Malanczuk. *Akehurst's Modern Introduction to International Law*. 7. ed. London, England: Routledge, 1997. p. 198.

returned to USA years later, in an exchange of spies.[54] Recently, in 2001, the forced landing of another American spy aircraft in China caused major turmoil.[55]

After indicating that the 1944 Convention shall be applicable to activities performed by civil aircraft, and not to state aircraft (article 3),[56] it is established in the Chicago treaty that, in relation to non-schedule flights (article 5), "each contracting State agrees that all aircraft of the other contracting States, being aircraft not engaged in scheduled international air services shall have the right, subject to the observance of the terms of this Convention, to make flights into or in transit non-stop across its territory and to make stops for non-traffic purposes without the necessity of obtaining prior permission, and subject to the right of the State flown over to require landing." But such article reserves an important limitation: "each contracting State nevertheless reserves the right, for reasons of safety of flight, to require aircraft desiring to proceed over regions which are inaccessible or without adequate air navigation facilities to follow prescribed routes, or to obtain special permission for such flights."

As far as scheduled flights are concerned, that is those conducted by companies constituted to exploit commercial aviation, article 6 is applicable: "no scheduled international air service may be operated over or into the territory of a contracting State, except with the special permission or other authorization of that State, and in accordance with the terms of such permission or authorization".

The study of the above mentioned dispositions allows one to observe that the freedom of over flight is restricted, by article 5 of the Chicago Convention, to civil aircraft used for non-scheduled flights, belonging to State Parties, and in accordance to routes previously traced by the territorial State. The State being flown over may even restrict or prohibit flights above sensitive facilities for security reasons

[54] "Em 4 de julho de 1956, um novo avião-espião americano, o U-2, fez seu voo inaugural diretamente sobre Moscou e Leningrado, tirando excelentes fotografias de uma altura bem acima das possibilidades dos aviões de caça e mísseis antiaéreos soviéticos. (...) Os voos continuaram em intervalos regulares durante os quatro anos seguintes. Os russos, que conseguiam detectá-los no radar mas não tinham como abatê-los, limitavam-se a protestos superficiais, não querendo propagar sua incapacidade de controlar seu espaço aéreo. Os americanos, cientes de que os voos violavam a lei internacional, ficavam calados enquanto exploravam essa mina de informações para a inteligência". John Lewis Gaddis. *História da Guerra Fria.* Rio de Janeiro, Brazil: Nova Fronteira, 2006. p. 70.

[55] William R. Slomanson. *Fundamental Perspectives on International Law.* 4. ed. Belmont, USA: Wadsworth/Thomson Learning, 2003. p. 283.

[56] "(a) This Convention shall be applicable only to civil aircraft, and shall not be applicable to state aircraft. (b) Aircraft used in military, customs and police services shall be deemed to be state aircraft. (c) No state aircraft of a contracting State shall fly over the territory of another State or land thereon without authorization by special agreement or otherwise, and in accordance with the terms thereof. (d) The contracting States undertake, when issuing regulations for their state aircraft, that they will have due regard for the safety of navigation of civil aircraft."

(article 9),[57] as well as demanding prior authorization, as long as it is required without distinction between foreign and national aircraft (article 5). Such requirements are even more strict in relation to commercial air companies that perform scheduled flights, which may only operate in other States if necessary authorization is granted for over flight or landing (article 6).

The possibility of closing national airspace, as provided by article 9, (b), highlights the strategic importance of that part of the territory to every State. But it must be recalled that such alternative is restricted to exceptional circumstances or emergencies, or in the interest of public safety, as determined by the territorial State. Such closure may be restricted to certain areas or even encompass the whole country. In case of violation by foreign aircraft, the territorial State shall hold the right to demand immediate landing. Mentioned Hildebrando Accioly: "if an aircraft breaches one of the recognized rights of the overflown State, it is natural that such State may perform over it some sanction".[58]

There are relevant examples of partial or complete closure of national air space based on article 9 of the Chicago Convention: for instance, Spain, in April 1967, after Great Britain established a flying zone limitation around Gibraltar, forbade any flights over or around the Algeciras Bay for national security reasons.[59] Similar action was adopted by the USA government after the 9-11 terrorist attacks to the World Trade Center Towers in New York; for long hours, only military aircraft and the presidential airplane, *Air Force One*, were authorized to flight over American territory.[60]

Invasion of air space by civil aircraft often leads to violent reprisals by the territorial State based on the logic of national defense, causing the death of passengers,

[57] "(a) Each contracting State may, for reasons of military necessity or public safety, restrict or prohibit uniformly the aircraft of other States from flying over certain areas of its territory, provided that no distinction in this respect is made between the aircraft of the State whose territory is involved, engaged in international scheduled airline services, and the aircraft of the other contracting States likewise engaged. Such prohibited areas shall be of reasonable extent and location so as not to interfere unnecessarily with air navigation. Descriptions of such prohibited areas in the territory of a contracting State, as well as any subsequent alterations therein, shall be communicated as soon as possible to the other contracting States and to the International Civil Aviation Organization. (b) Each contracting State reserves also the right, in exceptional circumstances or during a period of emergency, or in the interest of public safety, and with immediate effect, temporarily to restrict or prohibit flying over the whole or any part of its territory, on condition that such restriction or prohibition shall be applicable without distinction of nationality to aircraft of all other States. (c) Each contracting State, under such regulations as it may prescribe, may require any aircraft entering the areas contemplated in subparagraphs (a) or (b) above to effect a landing as soon as practicable thereafter at some designated airport within its territory."

[58] "Se uma aeronave viola algum dos direitos reconhecidos ao Estado sobre o qual voa, é natural que o Estado sobrevoado exerça sobre ela alguma sanção." Hildebrando Accioly. *Tratado de Direito Internacional Público*. Vol. II. 3. ed. São Paulo, Brazil: Quartier Latin, 2009. p. 336.

[59] I. H. Diedericks-Verschoor. An Introduction to Air Law. 5. ed. Deventer, the Netherlands: Kluwer, 1993. p. 9.

[60] Anthony Aust. *Handbook of International Law*. Cambridge, England: Cambridge University Press, 2005. p. 353.

as indicated by Malcolm N. Shaw[61]: in 1955, a civil aircraft of the El Al Israel Airlines was downed by Bulgarian air fighters for violation of national air space, while in 1973 it was Israel that attacked a foreign civil aircraft, of the Libya Airlines company. The most controversial case involved the assault to a Korean Airlines commercial aircraft by the Soviet air force in 1983, claiming the life of all 269 people on board. Records collected latter determined that the aircraft, during a scheduled flight, lost its route due to technical problems, leading it into restricted areas of Soviet air space. The incident caused a prompt reaction by ICAO, which amended article 3 of the Chicago Convention by the Montreal Protocol of 1984, as well as revising the treaty's Annex II, regarding rules of air. These rule changes were meant in order to avoid the use of arms (not force, it must be stressed) against civil aircraft, in case of interception.[62]

In the case of violation of air space by terrorist attack using hijacked civil aircraft, the territorial State shall always have the right to defensive measures in accordance to article 51 of the UN Charter.[63] Malcolm N. Shaw concluded that "the intercepting forces may have to take action, within the parameters of proportionality, to forestall that threat and force may be necessary."[64]

I. H. Diedericks-Verschoordraws attention to the fact that article 9 of the Chicago Convention "highlights tendency of States to put their own interests first in certain circumstances, like political aspirations, military necessity or public safety. (….) Article 9, like Article 1, clearly and unmistakably reflects both the old principle of the sovereignty of a State over the airspace above its territory and the priority given by States to safeguarding their own interests".[65] In conclusion: the national air space shall only be opened for overflight whenever the territorial State (even if party to the Chicago Convention) considers that its sovereign interests were, are and will be guaranteed. Paulo Borba Casella indicated that the World Court already recognized such national prerogative: "the International Court of Justice affirmed, in its judgment of the case regarding military and paramilitary activities in Nicaragua and against it (1986), that the non-authorized overflight constitutes a breach to the principle of territorial sovereignty of States."[66]

A valuable observation must be advanced on this point, as explained by J. E. S. Fawcett: "State sovereignty over the superjacent air space, which is an established rule of international law, is therefore not dominium over it as a volume of air

[61] Malcolm N. Shaw. *International Law*. 5. ed. Cambridge, England: 2003. p. 473/479.

[62] Nguyen Quoc Dinh, Patrick Daillier e Alain Pellet. *Direito Internacional Público*. 2. ed. Lisbon, Portugal: Fundação Calouste Gulbekian, 2003. p. 1275.

[63] Paulo Borba Casella. *Direito Internacional, Terrorismo e Aviação Civil*. São Paulo, Brazil: Quartier Latin, 2006.

[64] Malcolm N. Shaw. *International Law*. 5. ed. Cambridge, England: 2003. p. 476.

[65] I. H. Diedericks-Verschoor. An Introduction to Air Law. 5. ed. Deventer, the Netherlands: Kluwer, 1993. p. 14.

[66] "A Corte Internacional de Justiça afirmou, no julgamento do caso das atividades militares e paramilitares na Nicarágua e contra esta (1986), que o sobrevoo não autorizado constitui ofensa ao princípio da soberania territorial dos Estados." Paulo Borba Casella. *Direito Internacional dos Espaços*. São Paulo, Brazil: Atlas, 2009. p. 509.

or of space, but rather imperium over activities or objectives above its territory, where 'above' is a defined geometrical relation between the activity or object and the territory; that is to say, the subjacent State alone can determine what activities may take place there and it has a right, in the absence of special agreement, to exclude any intruder, whether that intruder be a spacecraft, aircraft, radio wave, or noxious fumes."[67]

As previously mentioned, the principle of national sovereignty of States over their air space was adopted by municipal legislation, becoming a general principle of law after the Paris Convention of 1919. Brazil has a long track of the use of such rule. It was addressed in several pieces of legislation: Ruling Annexed to Decree 16,983, of 07.22.1925, Decree 20,914, of 01.06.1932, Decree-Law 483, of 06.08.1938, Decree-Law 32, of 11.18.1966 (and later revisions). The present Brazilian Aeronautical Code, Law 7,565, of 12.19.1986, provides in article 11 that "Brazil has complete and exclusive sovereignty over the air space above its territory and its territorial sea."[68] José da Silva Pacheco referred to laws of other countries in the same sense, including, for instance, article 3 of the *Codice della Navigazione de Italia*, article 1 of the Aeronautical Code of Uruguay, article 202 of the Aeronautical Code of Argentina, articles 1 e 2 of the *Ley de Aviación Civil da Colombia*, article 1 of the General Regulation of Air Navigation of Cuba, article 22 of the *Ley sobre Navegación Aérea de Chile*, and article 1 of the Spanish Law of Air Navigation.[69]

As far as telecommunications are concerned, using "hertzian waves", the principle of national sovereignty over the air space is applied in a very similar way to Air Law. Such reasoning was originally discussed as early as 1865, when the pioneer International Telegraph Union (ITU) was conceived, and was incorporated on several conventions about radiotelegraphy and telecommunications, including the ones of Berlin (1906), London (1912), Washington (1927), Madrid (1932) and Atlantic City (1947). It must be recalled that the International Telegraph Union would later be merged with the International Radiotelegraph Union (IRU)[70] into a new international organization, the International Telecommunication Union (ITU), in 1947.[71] In this new body, a complex international system of rules developed that faced, from the beginning, the challenge of conciliating the sovereignty of States with the growing importance of telecommunications to the economic, cultural and social development of States.[72]

[67] J. E. S. Fawcett. *International Law and the Uses of Outer Space.* Manchester, England: University Press, 1969. p. 19.

[68] "O Brasil exerce completa e exclusiva soberania sobre o espaço aéreo acima de seu território e mar territorial."

[69] José da Silva Pacheco. *Comentários ao Código Brasileiro de Aeronáutico.* 4. ed. Rio de Janeiro, Brazil: Forense, 2006. p. 39.

[70] Created at the Berlin Conference of 1906.

[71] ITU Constitution, Atlantic City, 1947.

[72] Colin D. Long. *Telecommunications Law and Practice.* 2. ed.. London, England: Sweet & Maxwell, 1995. p. 206/210.

Indeed, David I. Fischer proposes that "freedom of broadcasting" has achieved customary status throughout the years: "the early theories that radio waves would be subjected to the same principle of sovereignty as aircraft in national air space were not sustainable in the long run, as evidenced by the fact that states have never claimed sovereignty over radio waves. Early theoretical attempts to incorporate radio waves into the existing regime of air space were thus abandoned. In its place came the development of a rather pragmatic approach which lead to a functional 'freedom of broadcast.'"[73] In accordance with Fischer, such freedom would not be applicable regarding propaganda broadcasts which endanger state security and order, pirate broadcasts and broadcasts causing harmful interference to the broadcasting activities of other states in violation of the technical regulations of the ITU.[74]

In conclusion, States exercise complete and exclusive sovereignty over their respective air spaces, especially regarding aviation. The same is applicable to telecommunications, taking into consideration practical limitations. Nevertheless, no international treaty that contemplated the topic, including the Chicago Convention of 1944, has delimited the vertical limit of national sovereignty. That would eventually constitute a problem to International Law.

Outer Space: International Territory

Space Law was conceived based on the prohibition of national appropriation, inserting outer space into existing international law regarding the high seas and the Antarctic continent. Accordingly, no State has the right, by any means, to claim Earth orbits or celestial bodies under its sovereignty, as if they were *res nullius*, opened to traditional methods of annexation, for instance discovery or occupation. Outer space constitutes *res communis omnium*, therefore free for appropriation even by individuals, that are subjected to the jurisdiction of their respective territorial or national State.[75] As noted by José Monserrat Filho, due to the development of Space Law, "it became illegal any idea of colonizing outer space and celestial bodies, as seen in the past in relation to vast regions of the planet."[76]

[73] David I. Fischer. *Prior Consent to International Direct Satellite Broadcasting*. Dordrecht, the Netherlands: Martinus Nijhoff, 1990. p. 159.

[74] David I. Fischer. *Prior Consent to International Direct Satellite Broadcasting*. Dordrecht, the Netherlands: Martinus Nijhoff, 1990. p. 160.

[75] Manfred Lachs affirmed that the outer space is not a thing, but a place, so that the terms *res communis* and *res extra commercium* must be understood under the specific context of Space Law: "Outer space and celestial bodies are to be viewed as spheres of States' activities; as an environment subjected to a special legal régime and enjoying that particular protection of the law. This determines the relationship between States and the new dimension." *The Law of Outer Space*. Leiden, the Netherlands: Martinus Nijhoff, 2010. p. 46.

[76] "[com o desenvolvimento do Direito Espacial] tornou-se ilegal qualquer ideia de colonizar o Espaço e os corpos celestes, como se viu no passado em vastas regiões do planeta." José Monserrat Filho. *Direito e Política na Era Espacial*. Rio de Janeiro, Brazil: Vieira & Lent, 2007. p. 32.

As explained by Gyula Gál, "outer space is, morally speaking, the common property of all mankind, because the Space Treaty regards the space exploration and use of the cosmic space as 'the joint venture of all mankind' declaring, as it does, the right of all States, without discrimination and based on sovereign equality, to carry on space activities. This is the meaning of the term *res communis omnium* and nothing else."[77]

The concept of "international territories" represents an important feature of post-modern International Law, displaying a relevant qualitative change in the Law of Spaces. As taught by Paulo Borba Casella:

> In accordance with the traditional view, strictly interstate wise, the international spaces, as zones beyond reach of State sovereignty, would be similar to a res nullius. As such, could be considered possible of appropriation and exploitation, indiscriminately, and that still remained, without substantive changes, until the postmodern context. Only during the last few decades has a different direction gained momentum, tending to the recognition of collective interests, regarding the 'whole mankind', and the affirmation of shared condition of this group of spaces, not subjected to the jurisdiction of States, and of the common interests in relation to them. (…) The international spaces, from no-man's land (res nullius) have become considered common good (*res communis*).[78]

The inaugural mission performed by the Soviet spacecraft Sputnik I, in 1957, constituted the turning point for consolidation of an international custom regarding the nonexistence of State sovereignty over outer space. Although Sputnik I could be easily tracked during its orbital path, no State presented formal protest based on the violation of its air space by the USSR. On the contrary, congratulatory compliments were offered throughout the globe for the successful mission, including by the USA, who used the opportunity to plead for the liberty of outer space.[79] Such general practice configured the necessary *opinio iuris* to justify the development of a consuetudinary norm. Since during the subsequent space

[77] Gyula Gál. *Space Law*. Leiden, the Netherlands: A. W. Sijthoff, 1969. p. 123/124.

[78] "Segundo a visão tradicional, estritamente interestatal, os espaços internacionais, enquanto zonas situadas além do alcance da soberania de cada um dos estados, se equiparavam a res nullius. Como tais, eram considerados passíveis de apropriação e de exploração, indiscriminadas, e isso permanece, sem substanciais alterações, até o contexto pós-moderno. Somente no curso das últimas décadas se põe sentido diverso, tendente ao reconhecimento de interesses coletivos, atinentes a 'toda a humanidade', a afirmação de condição compartilhada do conjunto dos espaços, não-sujeitos à jurisdição dos estados, e do interesse comum em relação a estes. (…) Os espaços internacionais, de terra de ninguém (res nullius) passam a ser considerados bem comum *(res communis)*". Paulo Borba Casella. *Direito Internacional dos Espaços*. São Paulo, Brazil: Atlas, 2009. p. 566/567.

[79] Paul G. Dembling e Daniel M. Arons make reference to a declaration of the American president Dwight D. Eisenhower at UN General Assembly in 1960, where he categorically affirmed that the USA understood that celestial bodies could not be subjected to national claims. "The Evolution of the Outer Space Treaty", Journal of Air, Law and Commerce, 419, 1967. *Apud* Glenn H. Reynolds and Robert P. Merges. *Outer Space: Problems of Law and Policy*. 2. ed. Boulder, USA: Western Press, 1998. p. 70.

missions the international opinion did not change, a *consuetudo* has been consolidated, supporting the new rule.[80]

In the aftermath of the inaugural space missions, it was the general opinion that innovative rules should be provided for the use and exploration of this new realm. As emphasized by Henry A. Wassenbergh, "the rule of law is necessary to keep the peace and provide security by balancing the various interests of those aspiring to be active in the same area, i.e. in outer space."[81] Discussions regarding the legal regime applicable to outer space gained momentum in the UN General Assembly from 1957, and resolutions were approved in rapid succession. Although representing an example of *soft law*, therefore being non-binding *per se*, those instruments consolidated or initiated international customs in relation to a territory which, by that time, had not been properly considered by International Law.[82]

The first of those resolutions, identified as 1348 (XIII), approved just a month after Sputnik I's launching, stated that space missions should be for peaceful uses only and, due to that, States had to maintain a clear channel for communications about outer space.[83] The intention of that instrument was crystal clear: avoid greed among nations and wars in the new territory of space.[84]

During the following sessions, the legal nature of outer space received progressively more attention. As observed by Valnora Leister, the USA proposed the creation of an *ad hoc* committee for discussion of questions related to outer space, based on an autonomous context in relation to disarmament issues.[85] Such body later became permanent, under the title of United Nations Committee on the Peaceful Uses of Outer Space (UNCOPUOS), through Resolution 1472 (XIV), of

---

[80] "Conforme a tradição, a unanimidade da doutrina internacionalista e inúmeros precedentes de tribunais internacionais, para que um comportamento comissivo ou omissivo seja considerado como um costume jurídico internacional, torna-se necessária a presença de dois elementos constitutivos: (a) um elemento material, a 'consuetudo', ou seja, uma prática reiterada de comportamentos, que, no início de sua formação, pode ser um simples uso ou prática; e (b) um elemento psicológico, ou subjetivo, a 'opinio júris vel necessitatis', ou seja, a certeza de que tais comportamentos são obrigatórios, em virtude de representarem valores essenciais e exigíveis de todos os agentes da comunidade dos Estados." Guido Fernando Silva Soares. *Curso de Direito Internacional Público*. São Paulo, Brazil: Atlas, 2002. p. 82.

[81] Henri Wassenbergh. *Principles of Outer Space in Hindsight*. Dordrecht, the Netherlands: Kluwer Academic Publishers, 1991. p. 16.

[82] Bin Cheng. *Studies on International Space Law*. Oxford, England: Clarendon Pr, 1998. p. 125/148.

[83] http://www.oosa.unvienna.org/oosa/SpaceLaw/gares/html/gares_13_1348.html, access on 01.10.2011.

[84] In 1959, the ad hoc study committee created by Resolution 1348 (XIII) presented report, referred by Ogunsola O. Ogunbanwo: "without necessarily going into details of the report of the Ad Hoc Committee, one or two observations will be necessary. First, the Committee considered that 'as a matter of principle the UN Charter and the Statute of the International Court of Justice were not limited in their operation to the confines of Earth'." *International Law and Outer Space Activities*. The Hague, the Netherlands: Martinus Nijhoff, 1975. p. 12.

[85] Valnora Leister. "O Comitê para o Uso Pacífico do Espaço Exterior (COPUOS) da Organização das Nações Unidas (ONU)", *In*: Araminta Mercadante e José Carlos de Magalhães (orgs.) *Reflexões sobre os 60 Anos da ONU*. Ijuí, Brazil: Unijuí, 2005. p. 400.

December 12th, 1959.[86] Deliberations in this new UN forum only started in 1961, when the Soviet proposal for decisions to be taken based solely in consensus was finally accepted.[87] To strengthen the quality of debates, two additional features were added: the Scientific and Technic Subcommittee and the Legal Subcommittee.[88] Subsequently, the United Nations Office for Outer Space Affairs (UNOOSA) was established as the executive office of UNCOPUOS.

Through this new body of the United Nations, international delegations, led by the USA and the USSR, conceived treaty rules regarding Space Law in a fast paced burst of legislative activity that can be even considered revolutionary. But prior to that, some fundamental principles had to be clarified. The newly born Legal Subcommittee assumed this responsibility. The UNGA Resolution 1721 (XVI), of December 20th, 1961,[89] represented the effort of UNCOPUOS to set aside any statements that outer space could eventually constitute a form of *res nullius*, therefore subjected to sovereignty claims. In its brief text, part "A", said Resolution reaffirmed the applicability of International Law, including the UN Charter, to outer space.[90] Next, it clearly stated that "outer space and celestial bodies are free for exploration and use by all States in conformity with international law and are not subject to national appropriation."[91]

The rule presented by Resolution 1721 (XVI), in accordance to the opinion presented by C. Wilfred Jenks, allowed the understanding that the principle of non-appropriation of outer space was already effectively consolidated in International Law by that time. The renowned lawyer clarified that, since States bear international responsibility for national activities in space, "it follows that what is forbidden to a State is not permitted to a chartered company created by a State or to one of its nationals acting as a private adventurer. (…) Only as regards a possible appropriation by the United Nations acting on behalf of the world community as a whole can the matter be regarded as an open one for the future".[92]

---

[86] http://www.oosa.unvienna.org/oosa/SpaceLaw/gares/html/gares_14_1472.html, access on 02.10.2011.

[87] "The enlargement of COPUOS and the strengthening of the Soviet bloc component in its composition was not the only concession made to the Soviet Union in order to secure its co-operation. From the start, being in the minority, the Soviet Union had wanted the unanimity rule to be applied in COPUOS, instead of the majority rule applicable to all subordinate organs of the United Nations General Assembly". Bin Cheng. *Studies on International Space Law*. Oxford, England: Clarendon Pr, 1998. p. 163.

[88] MATTE, Nicolas Mateesco. *Aerospace Law: from Scientific Exploration to Commercial Utilization*. Toronto, Canada: Carswell, 1977. p. 21/30.

[89] http://www.oosa.unvienna.org/oosa/SpaceLaw/gares/html/gares_16_1721.html#secte, access in 02.10.2011.

[90] Francys Lyall e Paul B. Laursen. *Space Law: a Treatise*. Farnham, England: Ashgate, 2009. p. 48.

[91] "1. Commends to States for their guidance in the exploration and use of outer space the following principles: (a) International law, including the Chapter of the United Nations, applies to outer space and celestial bodies; (b) Outer space and celestial bodies are free for exploration and use by all States in conformity with international law and are not subject to national appropriation; (…)."

[92] C. Wilfred Jenks. *Space Law*. New York, USA: Frederick A. Praeger, 1965. p. 201.

The internationally well-received Resolution 1721 (XVI) was in accordance with the speech presented by the American president John F. Kennedy at the UNGA in 1961. As indicated by Bin Cheng, it seemed to represent a binding rule to delegations of both the USA and the USSR, whether due to the fact that it consolidated a recognized international custom or because it was supported by unanimous international approval.[93]

During the following sessions, the debates at the Legal Subcommittee allowed the identification of basic rules of Space Law, but what kind of international instrument should be adopted for future regulation remained to be discussed. As mentioned by C. Wilfred Jenks, at the time, no consensus had been achieved to contemplate the rules through a comprehensive Space Law treaty, so the alternative was to issue another UNGA resolution, encompassing already accepted values.[94]

Titled the "Declaration of Legal Principles Governing the Activities of States in the Exploration and Use of Outer Space", the resolution was approved by the UNGA in December 13th, 1963, under the number 1962 (XVIII).[95] The document proclaimed the principles according to which States should conduct their space activities. Among these the two first principles stood out, proclaiming: "the exploration and use of outer space shall be carried on for the benefit and in the interests of all mankind" and "outer space and celestial bodies are free for exploration and use by all States on a basis of equality and in accordance with international law."

Regarding the legal status of outer space, the Declaration of Legal Principles stated, as its third principle, that outer space, including celestial bodies, could not be subject to national appropriation of any kind.[96] Hence, neither the continuous use of a certain orbital position nor the creation of permanent settlement in a celestial body could be used to justify any national claim. The text, approved unanimously at the UNCOPUOS legal subcommittee, was achieved due to a relevant compromise obtained by the USA and USSR, then the only space powers. Based on the success of the "Treaty Banning Nuclear Weapon Tests in the Atmosphere, in Outer Space and Under Water", signed in 1963,[97] their delegates were capable of establishing some important common opinions in relation to International Law, an impressive accomplishment during the Cold War.[98]

[93] Bin Cheng. *Studies on International Space Law*. Oxford, England: Clarendon Pr, 1998. p. 127.

[94] "The Legal Sub-Committee failed to break the deadlock which had frustrated its work but reported some progress. On the question of general principles governing the activities of States relating to the exploration and use of outer space, agreement was reached that they should take the form of a declaration, some delegations favoring a treaty-type document and others a General Assembly resolution." C. Wilfred Jenks. *Space Law*. New York, USA: Frederick A. Praeger, 1965. p. 63.

[95] http://www.unoosa.org/pdf/gares/ARES_18_1962E.pdf, access on 02.10.2011.

[96] "Outer space and celestial bodies are not subject to national appropriation by claim of sovereignty, by means of use or occupation, or by any other means."

[97] Later it was signed the Comprehensive Nuclear Test Ban Treaty, on September 10th, 1996, in New York, USA. Francys Lyall e Paul B. Laursen. *Space Law: a Treatise*. Farnham, England: Ashgate, 2009. p. 56.

[98] Bin Cheng. *Studies on International Space Law*. Oxford, England: Clarendon Pr, 1998. p. 153.

The Resolution 1962 (XVIII) consolidated several rules of Space Law: not only the prohibition of national claims, but also the need of peaceful cooperation and the international responsibility of States for national space activities. Also, the registration of space objects and the rights of astronauts were introduced in a rather inventive way. The document exercised definitive influence in International Law, and its provisions would later be incorporated in successive Space Law treaties.[99]

When the 43 members of UNCOPUOS approved, without dissent, the definitive text of the "Treaty on Principles Governing Activities of States in the Exploration and Use of Outer Space, Including the Moon and Other Celestial Bodies", international customs then already effective were reaffirmed by a binding written instrument. The wording made constant reference to past resolutions of the UNGA, with minor modifications. Therefore, the "Space Treaty", as it became knew worldwide, opened for signature on January 27th, 1967 in London, Moscow and Washington, was laid upon concepts and principles that had been recognized as of a fundamental character by the UNCOPUOS legal subcommittee. The extraordinary speed it was concluded in could only be achieved through the combination of American and Soviet drafts, presented during the conception of governmental projects for exploration of the Moon.[100]

It is important to verify, as addressed by Julia Neumann in her study regarding interpretation of the Space Treaty, that such agreement did not include a definition of outer space, its subject matter *per se*. To Neumann, "a certain area far enough from the Earth is clearly understood from the ordinary meaning of the term as 'outer space'."[101]

---

[99] C. Wilfred Jenks. *Space Law*. New York, USA: Frederick A. Praeger, 1965. p. 184/186.

[100] Bin Cheng presents a brief historical record of the Space Treaty creation: "The impetus behind the conclusion of this treaty was the successful 'soft' landing by the Soviet Union of its automatic station Luna IX on the moon on 3 February 1966, after three failures the previous years. Speaking from his ranch in Texas on 7 May 1966, President Johnson announced that the United States, which had hitherto opposed the development of international space law through treaties, would seek a treaty through the United Nations to prevent any nation from claiming sovereignty over the moon or any other celestial bodies and that the exploration thereof would be for peaceful purposes only. Consultations with the Soviet Union took place on 11 May when an outline of twelve points which, in the opinion of the United States, should be in a 'celestial bodies treaty' was handed to the latter. The response of the Soviet Union was prompt. On 30 May, it requested the 'Conclusion of an international agreement on legal principles governing then activities of States in the exploration and conquests of the Moon and other celestial bodies' to be included in the Agenda of the 21st session of the General Assembly. The text of a Soviet draft treaty followed on 16 June 1966. Whilst both drafts and their successive amendments were considered in the Legal Sub-Committee of COPUOS, in the course of which the United States agreed to enlarge the scope of the treaty to include also outer space, the really crucial discussions took place directly between the two space powers until they were able to present an agreed text in the form of a 43-power sponsored draft resolution to the First Committee of the General Assembly on 15 December 1966 for it to be adopted unanimously four days later by the General Assembly itself on 19 December 1966 in resolution 2222 (XXI). The actual treaty was then opened for signature on 27 January 1967 simultaneously in London, Moscow and Washington DC, the capitals of the three depository States." *Studies on International Space Law*. Oxford, England: Clarendon Pr, 1998. p. 156.

[101] Julia Neumann. "An Interpretation of the Outer Space Treaty after 40 years". *Proceedings of the Fiftieth Colloquium on the Law of Outer Space*. IISL, Hyderabad, 2007.

In accordance with the Space Treaty, outer space is the province of mankind.[102] Reviewing such provision, Stephan HOBE indicated that the article first "had as its aim an approach that disregards national interests and takes up the interests of all mankind. Outer space shall not be used in the interest of nation states alone, but shall be used and explored in the interests of all humankind".[103]

The following provision clearly prohibited any form of sovereign annexation in relation to the territory in question: "outer space, including the Moon and other celestial bodies, is not subject to national appropriation by claim of sovereignty, by means of use or occupation, or by any other means."[104] As remarked by I. H. Ph. Diederiks-Verschoor, "the formula epitomizes the great difference between air law and space law, with the former still embracing national sovereignty; while in the later claims of sovereignty are banned. (…) the ban on sovereignty remains dearly expressed as a fundamental factor in space law, and it must be seen as constituting an absolute legal barrier in the realization of every kind of space activity."[105]

According to Valérie Kayser, that rule constitutes example of *jus cogens*, applicable even to States that are not party to the Space Treaty. The attorney identified an interesting consequence of this, especially in comparison to what happened in relation to Air Law: "by depriving States of the attribute of sovereignty in relation to outer space, these principles have also worked against the motivation of States for establishing a detailed regulatory framework, since such a framework is not needed to ensure their compliance with the international framework which governs their activities in the space sector, and is therefore not a prerequisite to the performance of any business related to space, in particular the launch operations."[106]

---

[102] Translated in the official Brazilian version as "incumbência da humanidade", a very questionable option, since it would mean in English something like "a duty of mankind".

[103] Stephen Hobe. "Outer Space as the Province of Mankind – an Assessment of 40 Years of Development". *Proceedings of the Fiftieth Colloquium on the Law of Outer Space*. IISL, Hyderabad, 2007.

[104] Paul G. Dembling reported the legislative process regarding article 2 of the Space Treaty: "The text of Article II, which prohibits national appropriation of outer space and celestial bodies, provoked only a few minutes of debate. The wording of the second sentence of Article I of the United States draft and the wording of Article II of the Soviet draft are almost identical. Agreement was reached on the final text when the Soviet delegate concurred with a suggestion by the United States delegate that the words 'and celestial bodies' in the Soviet draft be replaced by the words 'including the Moon and other celestial bodies' and other minor drafting change. Although there was some later criticism of the use of the word 'appropriation' for possible vagueness, the Soviet delegate had indicated, at a prior stage of discussions, that the term referred to the ban on assertion of national claims by way of any human activity in outer space or on the Moon or other celestial bodies." "Principles Governing the Activities of States in the Exploration and Use of Outer Space, Including the Moon and Other Celestial Bodies", *In:* Nandasiri Jasentuliyana and Roy S. K. Lee (eds.). *Manual on Space Law*. Volume I. New York, USA: Oceana, 1979. p. 11.

[105] I. H. Ph. Diederiks-Verschoor. *An Introduction to Space Law*. 2. ed. Dordrecht, the Netherlands: Kluwer Academic Publishers, 1999. p. 28.

[106] Valérie Kayser. *Launching Space Objects: Issues of Liability and Future Prospects*. Dordrecht, the Netherlands: Kluwer Academic Publishers, 2001. p. 27/28.

The international public concerns played a major part in the limitation of vertical sovereignty of States, concluded Henri A. Wassenburgh. To the author, a complex factor had a major relevance at the time: "it was essentially 'fear' which inspired States to declare the applicability of international law, including the Charter of the UN, to space activities. It was and is quite understandable that the States feared and still fear that space activities would interfere with the legal status quo on Earth, that space activities might change the balance of power on earth and thereby change the status quo as protected by international law and the Charter of the UN."[107] Due to these concerns, Article 3 of the Space Treaty provided that the activities in outer space shall be conducted "in accordance with international law, including the Charter of the United Nations, in the interest of maintaining international peace and security and promoting international cooperation and understanding."

The nonexistence of controversy in relation to the prohibition of appropriation by national claims in relation to outer space had an important consequence, as presented in a detailed study concluded by the University of Cologne regarding the Space Treaty: "it has been the established practice of the major space faring States that, unlike traditional notions of territory on Earth, no State could subject (any part of) outer space, including the moon and other celestial bodies, to its sovereign control, or regard it as part of its territory. This prohibition has served to protect outer space from national/exclusive colonization by States, which itself represents an important element in the peaceful uses of outer space."[108]

Usually considered the Magna Carta of Space Law,[109] the Space Treaty recently welcomed its 100th State as a signatory, and has served as a sound foundation for other treaties about outer space, all of them encompassing, in one way or another, the principle that the territory of space escapes from State sovereignty.[110]

Special reference amongst later treaties must be made to the "Convention on International Liability for Damage Caused by Space Objects", opened for signature on March 29th, 1972. This instrument identifies the international liability, regarding damage inflicted to third parties, by a State that launches or arranges the launching of a space object, or from whose territory or facility a space object is launched, considered therein as a "Launching State" (Article 1, "c"). The Liability Convention provides a double system of liability, depending on the place of damage: if caused on the surface of Earth or to an aircraft in flight, there shall be absolute liability (Article 2); on the other hand, if caused elsewhere than on the surface of Earth to another space object, liability can only be effective in case of fault (Article 3).[111]

---

[107] Henri A. Wassenberg. *Principles of Outer Space in Hindsight*. USA: Springer, 1991. p. 20.

[108] Steven Freeland and Ram Jakhu. "Article II", *In:* Stephan Hobe, Bernhard Schmidt-Tedd, Kai-Uwe Schrogl (eds.). *Cologne Commentary on Space Law.* Volume 1 – Outer Space Treaty. Cologne, Germany: Carl Heymanns, 2009. p. 53.

[109] Stephen Gorove. "Sources and Principles of Space Law", *In*: Nandasiri Jasentuliyana (coord.). *Space Law: Development and Scope*. Westport, USA: Praeger Publishers, 1992. p. 46.

[110] Nandasiri Jasentuliyana. *International Space Law and the United Nations*. The Hague, the Netherlands: Kluwer, 1999. p. 32/41.

[111] Olavo de O. Bittencourt Neto. *Direito Espacial Contemporâneo: Responsabilidade Internacional.* Curitiba, Brazil: Juruá, 2011. p. 80/89.

Therefore, the 1972 Convention is applicable to all space activities, considered not only as those that effectively take place in outer space, but also as the ones that, even though aiming to such territory, due to whatever reason cannot actually achieve it. Damage caused by the return of space objects to Earth is governed by the agreement as well, despite this damage not taking place in outer space. Anyhow, the grounds for liability are connected to the place of damage[112]: if it happens in a territory under national jurisdiction, be it on land, sea or air, the Launching State shall enjoy minimal causes for exclusion of liability (Article 2).[113] The situation is quite different for damage caused in outer space, where the classic fault based international liability of States is applicable, with all of its consequences to the parties (Article 3).[114]

The following space treaty, named "Convention on the Registration of Objects Launched into Outer Space", adopted by UNGA on November 12th, 1974 and opened for signature in New York on January 14th, 1975, establishes the obligation of the respective Launching State to entry, in a national registry that it shall create and keep updated, any space object "launched into Earth orbit or beyond" (Article 2).[115] The UN Secretary General shall also be informed of such activity, through the enrolment of certain data regarding the space object in question, featuring basic orbital parameters, including nodal period, inclination, apogee and perigee (Articles 3 and 4). As far as joint launches are concerned, each Launching State shall include it in its respective national register, but only one of them shall provide the international registration before the UN (Article 2, 2).[116]

Finally, the "Agreement Governing the Activities of States on the Moon and other Celestial Bodies", known as the "Moon Treaty", opened for signature in

[112] Olavo de O. Bittencourt Neto. *Responsabilidade Internacional dos Estados no Direito Espacial: Brasil como Estado Lançador.* São Paulo, Brazil: Faculdade de Direito da Universidade de São Paulo, Brazil, 2008. p. 72/82.

[113] "A launching State shall be absolutely liable to pay compensation for damage caused by its space object on the surface of the Earth or to aircraft in flight."

[114] "In the event of damage being caused elsewhere than on the surface of the Earth to a space object of one launching State or to persons or property on board such a space object by a space object of another launching State, the latter shall be liable only if the damage is due to its fault or the fault of persons for whom it is responsible."

[115] "1. When a space object is launched into Earth orbit or beyond, the launching State shall register the space object by means of an entry in an appropriate registry which it shall maintain. Each launching State shall inform the Secretary-General of the United Nations of the establishment of such a registry. 2. Where there are two or more launching States in respect of any such space object, they shall jointly determine which one of them shall register the object in accordance with paragraph 1 of this article, bearing in mind the provisions of article VIII of the Treaty on Principles Governing the Activities of States in the Exploration and Use of Outer Space, including the Moon and Other Celestial Bodies, and without prejudice to appropriate agreements concluded or to be concluded among the launching States on jurisdiction and control over the space object and over any personnel thereof. 3. The contents of each registry and the conditions under which it is maintained shall be determined by the State of registry concerned."

[116] Olavo de O. Bittencourt Neto. *Responsabilidade Internacional dos Estados no Direito Espacial: Brasil como Estado Lançador.* São Paulo, Brazil: Faculdade de Direito da Universidade de São Paulo, Brazil, 2008. p. 115.

New York on December 18th, 1979,[117] embodies the non-appropriation principle in its Article 11.[118] After identifying the Moon and its natural resources as "common heritage of mankind", the agreement provides: "the Moon is not subject to national appropriation by any claim of sovereignty, by means of use or occupation, or by any other means."

Such legal regime is explained by item 3 of said article 11, that deserves transcription: "neither the surface nor the subsurface of the Moon, nor any part thereof or natural resources in place, shall become property of any State, international intergovernmental or non-governmental organization, national organization or non-governmental entity or of any natural person. The placement of personnel, space vehicles, equipment, facilities, stations and installations on or below the surface of the Moon, including structures connected with its surface or subsurface, shall not create a right of ownership over the surface or the subsurface of the Moon or any areas thereof. The foregoing provisions are without prejudice to the international regime referred to in paragraph 5 of this article." The exploitation of natural resources from the Moon or other celestial bodies shall depend upon the establishment of an international regime, an endeavor that must be undertaken by the State-Parties as soon as those activities become feasible (Article 11, 5).

---

[117] Brazil is not a member-State.

[118] "1. The Moon and its natural resources are the common heritage of mankind, which finds its expression in the provisions of this Agreement, in particular in paragraph 5 of this article. 2. The Moon is not subject to national appropriation by any claim of sovereignty, by means of use or occupation, or by any other means. 3. Neither the surface nor the subsurface of the Moon, nor any part thereof or natural resources in place, shall become property of any State, international intergovernmental or non-governmental organization, national organization or non-governmental entity or of any natural person. The placement of personnel, space vehicles, equipment, facilities, stations and installations on or below the surface of the Moon, including structures connected with its surface or subsurface, shall not create a right of ownership over the surface or the subsurface of the Moon or any areas thereof. The foregoing provisions are without prejudice to the international regime referred to in paragraph 5 of this article. 4. States Parties have the right to exploration and use of the Moon without discrimination of any kind, on the basis of equality and in accordance with international law and the terms of this Agreement. 5. States Parties to this Agreement hereby undertake to establish an international regime, including appropriate procedures, to govern the exploitation of the natural resources of the Moon as such exploitation is about to become feasible. This provision shall be implemented in accordance with article 18 of this Agreement. 6. In order to facilitate the establishment of the international regime referred to in paragraph 5 of this article, States Parties shall inform the Secretary-General of the United Nations as well as the public and the international scientific community, to the greatest extent feasible and practicable, of any natural resources they may discover on the Moon. 7. The main purposes of the international regime to be established shall include: (a) The orderly and safe development of the natural resources of the Moon; (b) The rational management of those resources; (c) The expansion of opportunities in the use of those resources; (d) An equitable sharing by all States Parties in the benefits derived from those resources, whereby the interests and needs of the developing countries, as well as the efforts of those countries which have contributed either directly or indirectly to the exploration of the Moon, shall be given special consideration. 8. All the activities with respect to the natural resources of the Moon shall be carried out in a manner compatible with the purposes specified in paragraph 7 of this article and the provisions of article 6, paragraph 2, of this Agreement."

When studying Article 11 of the Moon Treaty, Peter P. C. Haanapel is emphatic: "the text is crystal-clear: there are no private ownership rights on the Moon or on other celestial bodies. This is wholly in keeping with the above-mentioned principle of non-appropriation. It applies also to orbits and trajectories around or to the Moon and other celestial bodies, and a fortiori to orbits around the Earth (...)."[119]

I. H. Ph. Diedericks-Verschoor notes that the use of the term "common heritage of mankind" by the Moon Treaty constituted an innovation in Space Law. Indeed, as previously mentioned, the Space Treaty establishes that the outer space constitute "province of mankind". In the opinion of Diedericks-Verschoor, there is a relevant difference between those two terms: "the 'province of mankind' must be identified as a general political principle with certain moral overtones, meant to govern rights and duties in outer space. Its legal substance, according to Article I [of the Outer Space Treaty], is international co-operation and use of outer space without discrimination of any states, and the duty to take into account the interests of other states. The scope of the term 'common heritage' is much more restricted in general terms, covering only the exploitation of the Moon's natural resources."[120]

Despite all the above, an orbital zone of high importance, extremely valuable for telecommunication, was subject to joint national claims by Equatorial States in a statement signed at the Colombian capital of Bogotá in 1976. Reference is made in this document to the geostationary orbit, located at 36,000 km of altitude from sea level. Geostationary orbit allows space objects to move at approximately the same speed of the Earth's rotation, so that they remain apparently still over a specific quadrant on the surface during their lifespan. Signed by several developing States, including Brazil (though Brazil later denied its terms and any binding effect regarding the declaration), this instrument failed to obtain recognition by the international community. Nowadays, the geostationary orbit is generally regarded as a limited natural resource that shall be used for the benefit of all mankind, operating on the principle of equitable access to all States based on their respective national needs.[121]

[119] Peter P. C. Haanappel. *The Law and Policy of Air Space and Outer Space: a Comparative Approach*. Dordrecht, the Netherlands: Kluwer Academic Publishers, 2003. p. 25.

[120] I. H. Ph. Diederiks-Verschoor. *An Introduction to Space Law*. 2. ed. Dordrecht, the Netherlands: Kluwer Academic Publishers, 1999. p. 52.

[121] Milton L. Smith. *International Regulation of Satellite Communication*. Dordrecht, the Netherlands: Martinus Nijhoff, 1990. p. 201/206.

Chapter 3
The Delimitation in Discussion

> International law is a normative order, and a normative order is a system of valid norms. Legal norms regulate human behavior, and human behavior takes place in time and space. Consequently, legal norms have relation to time and space. [Hans Kelsen (1881–1973)] Principles of International Law[1]

The evolution of Space Law during the 1960s and 1970s happened in great speed. During a truly golden age, the five relevant Space Law treaties were concluded in rapid succession: the "Space Treaty" of 1967[2]; the "Rescue Agreement" of 1968[3]; the "Liability Convention" of 1972[4]; the "Registration Convention" of 1975[5]; and the "Moon Agreement" of 1979.[6] Those treaties were, in general, well received by the international community,[7] and provided legal rules applicable to the outer space and space activities. Nevertheless, some topics still remain unsolved despite their relevance and the special attention given to them by legal experts. This is the case of the definition and delimitation of outer space, a question that even predates the conclusion of the Space Treaty.[8] In 1966, years after the first discussions among

[1] Hans Kelsen. *Principles of International Law.* 2. ed. New York, USA: Holt, Rinehart and Winston, 1967. p. 177.

[2] "Treaty on Principles Governing the Activities of States in the Exploration and Use of Outer Space, Including the Moon and Other Celestial Bodies."

[3] "Agreement on the Rescue of Astronauts, the Return of Astronauts and the Return of Objects Launched into Outer Space".

[4] "Convention on International Liability for Damage Caused by Space Objects".

[5] "Convention on Registration of Objects Launched into Outer Space".

[6] "Agreement Governing the Activities of States on the Moon and Other Celestial Bodies".

[7] Notably, the Moon Agreement has not been granted the same appraisal of the other Space Law treaties referred.

[8] Ogunsola O. Ogunbanwo. *International Law and Outer Space Activities.* The Hague, the Netherlands: Martinus Nijhoff, 1975. p. 50/58.

© Springer International Publishing Switzerland 2015

O.O. Bittencourt Neto, *Defining the Limits of Outer Space for Regulatory Purposes*, SpringerBriefs in Space Development,
DOI 10.1007/978-3-319-16685-8_3

scholars,[9] the subject was included in the agenda of the UNCOPUOS legal subcommittee, based on a French delegation proposal,[10] but no solution has yet been achieved.

The debates on the subject were sometimes based on abstract scientific criteria, while other perspectives followed arguable presumptions. Due to the impasse among delegations, and also especially due to the lack of current practical problems surrounding the topic, the issue lost its priority. It shall be recalled that the approval of international instruments at the UNCOPUOS depends upon the consensus of all member States; therefore, a single opposition is capable of delaying discussions indefinitely.[11] Therefore, though the delimitation of outer space remains to be discussed at every session of the UNCOPUOS, the political will capable to bring a definitive solution to this standoff, or even to withdraw the topic from the legal subcommittee agenda, is still lacking.

"Functionalists" Versus "Spatialists"

The definition and delimitation of outer space has provoked heated debates among legal theorists over the years. These discussions involved not only legal experts, but also scientists and diplomats. Arguments based on astrophysics and legal science were proposed in order to provide parameters between the upper atmosphere and space that were capable of surviving criticism. But the debate still remains. Due to the lack of uniformity on the proposals of delimitation, another position exists that claims it is useless to provide a boundary between air space and outer space. This approach gained enough momentum to uphold the approval of alternatives before the UNCOPUOS.

The long period of discussion consolidated positions in two theoretical approaches: the one of the "spatialists", that defend the delimitation of the boundary

[9] One of the first remarks regarding the problem of delimitation of outer space was presented by Stephen Gorove, who, at the First Colloquium on the Law of Outer Space, held in the Hague in 1958, clearly stated: "He who controls the Cosmic Space, rules not only the Earth, but the whole Universe". His paper, that included the first proposals presented by publicists to solve such a question, was named "On the Threshold of Space: Toward a Cosmic Law – Problems of the Upward Extent of Sovereignty". *Proceedings of the First Colloquium on the Law of Outer Space*. IISL, the Hague, 1957. p. 69/76.

[10] Resolution 2222 (XXI), of December 19th, 1966.

[11] "Os USA propuseram à ONU que os assuntos do espaço cósmico fossem separados da questão de desarmamento e que se estabelecesse um comitê 'ad hoc' para assuntos espaciais e este posicionamento foi aceito pela URSS. Seguindo esta filosofia, houve votos suficientes para estabelecer o Comitê dos Usos Pacíficos do Espaço Exterior – COPUOS, através da Resolução 1742 (XIV) de 12.12.1959 sobre 'Cooperação Internacional sobre os Usos do Espaço Exterior. Mas o COPUOS só iniciou suas atividades em 1961 quando a URSS propôs aumentar o número de participantes para 28 de acordo com a representação geográfica. A URSS solicitou que as decisões fossem tomadas por consenso com a participação equitativa de países do mundo (…)." Valnora Leister. "O Comitê para o Uso Pacífico do Espaço Exterior (COPUOS) da Organização das Nações Unidas", *In*: Araminta Mercadante e José Carlos de Magalhães (coord.). *Reflexões sobre os 60 Anos da ONU*. Ijuí, Brazil: Unijuí, 2005. p. 400.

between air space and outer space, and the other of the "functionalists", which affirm that such an effort is fruitless or even impossible to be successfully achieved.

Despite the occasional shift of opinion, there is an interesting coherence among member States of UNCOPUOS in relation to their positions regarding delimitation, at least since the 1970s. Let's not forget that the Space Law treaties relied upon a convergence of opinions between the USA and USSR to be concluded, since only the two superpowers had the capability to reach Earth orbit during the early years of the conquest of space. But the Cold War adversaries parted ways in relation to the topic of delimitation: while the Soviets defended the delimitation of the air space/outer space boundary based on an arbitrary and conventional altitude, the Americans presented themselves in favor to the functional approach. The end of the Cold War has not changed this scenario. Brazil has also held its position throughout the years, sustaining that only a common and clear delimitation can provide legal security to space activities.

The fundaments of the functionalist approach were traced by Nicolas Mateesco Matte back to works developed by M. Lemoine, who stated that Space Law should focus its attention to space navigation. Later, the theory was perfected by other important legal theorists, including Joseph Kroell, R. Houmbourg, Rolando Quadri and finally Mateesco Matte himself, in his famous work "Deux Frontières Invisible: de la Mer Territoriale a l'Air 'Territorial'", of 1965.[12]

In that important work, Mateesco Matte explained the fundaments of what he then called "a new functional theory", which eventually constituted the arguments of every functionalist delegation before the UNCOPUOS later on. First, he stated that if the astronautic endeavors demanded the incidental passage of a space object through the territory of another State, no matter at which altitude, no danger could be apprehended as far as security is concerned. Second, if any damage is caused to the overflown State, it should have the right to receive reparation. Third, if such territorial State had serious reasons to believe that the space object is dangerous or is employed for aggressive purposes, it should have the right of defend itself. The functionalist approach, according to the author, excluded the necessity of delimitating frontiers between air space and outer space.[13]

[12] Nicolas Mateesco Matte. *Deux Frontières Invisibles: De la Mer Territoriale à l'Air 'Territorial'*. Paris, France: Pedone, 1965. p. 215/221.

[13] "Si de telles techniques astronautiques exigent que l'appareil survole incidemment le territoire d'un autre Estat, quelque hauteur que ce soit, il n'y a là rien de répréhensible ni de condamnable au point de vue de la sécurité des Etats. Si dans la poursuite de cet objectif cosmique le bolide causait des dommages à l'Etat survolé, il lui devra réparation. C'est là une règle non contestable. Si l'Etat survolé a de bonnes et sérieuses raisons de croire que le cosmonef causera des dommages ou sert à des fins agressives, il a droit de s'en défendre avec les moyens appropriés et proportionnés au danger. Pour échapper à l'autorité nationale survolée, le bolide devra se qualifier d'un objectif cosmique, spatial ou astronautique; ou encore il devra servir à des fins de recherchesm explorations, transport, communications, ravitaillement dans le cosmos, etc. (...) En conclusion, nous constatons que si une telle théorie fonctionnelle était adptée avec les ajustements qui s'imposeront à l'occasion, il n'y aura alors plus de nécessité ni d'utilité de fixer des frontières à l'air et à l'espace" Nicolas Mateesco Matte. *Deux Frontières Invisibles: De la Mer Territoriale à l'Air 'Territorial'*. Paris, France: Pedone, 1965. p. 223 and 226.

One of the major functionalist arguments questions if there is any scientific basis to claim that the air space has, in fact, a natural vertical limit. As early as 1967, it was requested to the UNCOPUOS scientific and technical subcommittee to provide a study on possible criteria for definition and delimitation of outer space.[14] The report presented by the experts concluded that, at the time, it was not possible to identify scientific or technical criteria that could allow a precise and lasting definition of outer space.[15] After four decades, despite interesting scientific experiments conducted about the Earth's orbit natural arrangements, the situation does not seem to have changed, at least significantly.[16] Nevertheless, Pierre-Marie Dupuy affirms categorically: the delimitation of the boundary between air space and outer space, sooner or later, shall require a clear solution.[17]

In the next pages, the explanations of those who defend the functionalist approach will be presented, as well as the ones of those who sustain the spatialist approach, in order to provide a better comprehension of each position. Such groups, although encompassing positions defended by several States and scholars, have distinctive features that must be acknowledged. Among the members of the UNCOPUOS legal subcommittee, the terms functionalist and spatialist were eventually accepted, and their positions in relation to this issue are usually presented as pertaining to one of the mentioned approaches. Therefore, there is a reasonably clear antithesis, as far as discussions at the UN are concerned.[18]

The States that follow the functionalist approach claim that the delimitation of the boundary between air space and outer space is either unnecessary or impossible to be properly accomplished.[19] In accordance to their position, the legal regime applicable to space activities should be defined taking into consideration not the place where they take place, but their nature and purpose. Therefore, all space activities, i.e., those related to the use of outer space, should be ruled by Space Law in all circumstances.[20] The raison d'être of the mission would have priority over the considerations related to the object's actual geographic location at any given time.[21]

---

[14] A/AC.105/37, para. 18.

[15] A/AC.105/C.2/SR.102-104 e 107.

[16] Robert F. A. Goedhart. *The Never Ending Dispute: Delimitation of Air Space and Outer Space*. Paris, France: Frontières, 1996.

[17] "Une importante question demeure à résoudre et appellera tôt ou tard une solution claire: c'est celle de la definition de la frontière qui le sépare de l'espace aérien national et international". Pierre-Marie Dupuy. *Droit International Public*. 8. ed. Paris, France: Dalloz, 2006. p. 788.

[18] Nandasiri Jasentuliyana. *International Space Law and the United Nations*. The Hague, the Netherlands: Kluwer, 1998. p. 51/52.

[19] Bin Cheng. *Studies on International Space Law*. Oxford, England: Clarendon Pr, 1998. p. 389.

[20] In relation to the "functional approach", Manfred Lachs stated: "it laid less emphasis on the location of the line to be drawn, much more on the character of activities of States in outer space and the objectives they pursue." *The Law of Outer Space*. Leiden, the Netherlands: Martinus Nijhoff, 2010. p. 53.

[21] Among the publicists that supported the functionalist approach, are included N. M. Matte, F. B. Schick, D. Goedhuis, C. Chaumont, R. Quadri, Seara Vasquez, H. A. Wassembergh e J. P. Pancracio.

As observed by Francisco Ferreira de Almeida, the functional approach is less connected to a territory than to the activities that take place therein.[22]

As explained by one of its most renowned advocates, Gyula Gál, based on the functional character of Space Law, "the new legal domain, other than all earlier norms of human conduct, cannot be associated with any limited space (area, zone), but only with the character of activity under regulation. (…) Its recognition is a logical necessity following from the various aspects of space law."[23]

In accordance to such reasoning, a space object, from the moment that it is put in motion by a launching vehicle, should be ruled by Space Law treaties, and not by those related to Air Law, even during such time when they may be found on or just a few meters above sea level. The same consideration would be applicable to reentries: there could not be, therefore, violation of foreign air space during the return of a space object to Earth, irrespective of its trajectory.[24]

As indicated by Francys Lyall and Paul B. Laursen, the first mention of the functionalist approach seems to be the one presented by Myres S. McDougall and L. Lipson, in 1958, just 1 year after Sputnik I's successful mission.[25] In order to support his thesis, McDougal, in a paper co-written with Harold D. Lasswell and Ivan I. Vlasic published in the early 1960s, submitted that the delimitation of the boundary between air space and outer space was defended by various legal minds as the only alternative against chaos, anarchy and other horrors. But, as he then remarked, "it is more than irony that after several years of space exploration and many successful flights of spacecraft, all in the absence of established boundaries, no serious dispute has yet occurred between states".[26]

Other publicists contributed to identify the fundaments of the functionalist approach, including Robert Houmborg, who applied it to aeronautical activities.[27] Rolando Quadri supported the use of functionalist rationale in relation to space programs during his course at the Hague Academy of International Law in 1958, affirming that only the Launching State should have jurisdiction over its own space objects.[28] Similarly, Charles Chaumont sustained the application of Space Law to every object intended to explore outer space, from its launching, irrespective of the

[22] Francisco Ferreira de Almeida. *Direito Internacional Público*. 2. ed. Coimbra, Portugal: Coimbra Editora, 2003. p. 214.

[23] Gyula Gál. *Space Law*. Leiden, the Netherlands: A. W. Sijthoff, 1969. p. 106.

[24] I. H. Ph. Diederiks-Verschoor. *An Introduction to Air Law*. 5. ed. Dordrecht, the Netherlands: Kluwer Academic Publishers, 1993. p. 20.

[25] M. McDougall and L. Lipson. "Perspectives for a Law of Outer Space". Apud: Francys Lyall and Paul B. Laursen. *Space Law: a Treatise*. Farnham, England: Ashgate, 2009. p. 169.

[26] Myres S. McDougal, Harold D. Lasswell and Ivan I. Vlasic. *Law and Public Order in Space*. New Haven, USA: Yale University Press, 1963. p. 323.

[27] Robert Houmborg. "Etendue et limites du Droit Aérien". *Revue Générale de l'Air.* Paris, France, 1956. p. 140/145; "Droit Astronautique et Droit Aérien". *Revue Générale de l'Air*. Paris, France, 1958. p. 11.

[28] Rolando Quadri. *Droit International Cosmique*. Recueil des Cours, vol. 58, n. III, the Hague, the Netherlands: 1958. p. 560.

place that it may be found.[29] Emil Konstantinov solved the question of legal regimes applicable to the air space and outer space establishing a connection between the activity and its respective legal system; according to his reasoning, a similar inference should be used in relation to hybrid vehicles, therefore identifying the major objective of their mission.[30]

One of the most important supporters of the functionalist approach, the already mentioned renowned Nicolas Mateesco Matte, asserted that, in relation to Air Law and Space Law, there is only a functional sovereignty for the former and a functional liberty for the later:

> In law, a science of men and of realities, the concepts of responsibility, liberty and sovereignty can be conceived only in regard to precise and concrete functions rather than in the abstract sense, and the application of these concepts must be reasonable.[31]

It is important to stress that none of the five major treaties of Space Law define nor delimit outer space, not even the Space Treaty of 1967; by the same token, Air Law does not provide the vertical limit of State sovereignty in any of its instruments, as can be verified by studying the Chicago Convention of 1944.[32] But it is important to recall, as prominently pointed out by Carl Q. Christol, that this is not unusual or abnormal, as far as territorial treaties are concerned.[33]

The functionalists assert that the absence of a clear boundary has not created any problem in the international community, at least up to this moment, and it would be

---

[29] Charles Chaumont. *Le Droit de l'Espace*. Paris, France: Presses Universitaires de France, 1960. p. 51.

[30] Emil Kostantinov. "Some Aspects of the Spatial and Functional Delimitation between International Air and Space". *Proceedings of the Twenty-Sixth Colloquium on the Law of Outer Space*. IISL, Budapest, 1983. p. 81.

[31] Nicolas Mateesco Matte. *Aerospace Law*. Toronto, Canada: Carswell, 1969. p. 63.

[32] In a literal interpretation to Chicago Convention's article 1, the delegation from Chile defended at the United Nations, in 1958, that since no vertical limit to national air space were provided by treaty, it should be understood as raising ad infinitum. A/C.1/PV.982. The opinions of Paul de La Pradelle can be mentioned in the same sense ("Le Frontiére de l'Air". *Recueil de Cours*, II. The Hague, the Netherlands, 1954. p. 121), R. C. Hingorani ("La Souverainete sur l'Espace Extra-Atmosphérique". *Revue Générale de l'Air*, Vol. 20. Paris, France, 1957. p. 248) and C. Papathanassiou ("Refléxions sur les Probléms Juridiques Posés par le Vol à Haute Altitude et le Vol Cosmique". *Revue Hellénique de Droit International*, Vol. 11, n.3-4. Athens, Greece, 1958. p. 248). Such a position, usually referred to as the *usque ad coelum* theory, would later be denied by space law treaties, which provided it to be free from national claims of any sort, as per article 1 of the Space Treaty, 1967.

[33] "Valid and effective legal regimes exist respecting the ocean's territorial waters without either a treaty base or a basis in customary international law fixing a single and uniform measurement. At the current Third United Nations Conference on the Law of the Sea [of 1982] no single and uniform measurement has been established for the point where the deep seabed and ocean floor begin and where the limits of national jurisdiction ends. Despite a series of treaties beginning with the 1919 Paris Convention and culminating in the 1944 Chicago Convention, none of the treaty-based rules of international law applicable to air space have fixed its upper boundary. The 1959 Antarctica Convention did not prescribe the metes and bounds of the Continent insofar as the Convention applied to a spatial area." Carl Q. Christol. *Modern International Law of Outer Space*. New York, USA: Pergamon Press, 1982. p. 443.

impossible to predict when, or even if, such an effort for delimitation could become justified.[34] Indeed, it is reckoned by Nguyen Quoc Dinh, Patrick Daillier and Alain Pellet that the pragmatic functionalist approach in fact allows most of the problems to be solved.[35]

In a deep study presented as PhD thesis to the University of Kent, United Kingdom, Gbenga Oduntan explained the inner features of such position:

> The crux of the functional approach lies then in the nature of the activities displayed or to be displayed. Thus, there is no distinction between air flight and space flight as well as between aircraft and spacecraft. By virtue of this reasoning wherever space objects may be found to be in operation, outer space laws apply.[36]

It is often espoused by the functionalists, for instance Nicolas Mateesco Matte,[37] that there aren't scientific criteria capable of identifying, without margin of doubt, the vertical limit of air space. Due to that fact, in their view, any delimitation would have to rely on dubious technical idiosyncrasies, or on an arbitrarily levied altitude. Such was the consideration espoused by Gyúla Gál during the 26th Colloquium on the Law of Outer Space promoted by IISL in Budapest, 1983, when he concluded that the functionalist approach is indispensable for the development of space activities, since it allowed a greater freedom of action for nations that develop space programs.[38]

The American delegation to the UNCOPUOS has traditionally endorsed the "functionalist" approach, suggesting that the provision of a conventional delimitation would imply impediments to future activities of exploration and exploitation of outer space, especially the commercial ones, conducted by private companies.[39] That's also the understanding of Jerome Morenoff, who claimed attention to the potential risks for air transport commercial exploitation if a fixed standard of altitude is established for national sovereignty.[40]

The lack of treaty settling the air space/outer space boundary, added to the absence of protests regarding the eventual occurrence of space launchings crossing foreign air space, would have provided, according to S. B. Rosenfield, a tacit acceptance of the functionalist approach. Therefore, it has been argued that an

---

[34] S. Neil Hosenball Jefferson S. Hofgard. "Delimitation of Air Space and Outer Space: Is a Boundary Needed Now". *University of Colorado Law Review*. Vol. 57. Boulder, USA, 1986. p. 892.

[35] Nguyen Quoc Dinh; Alain Pellet; Patrick Daillier. *Direito Internacional Público*. Lisbon, Portugal: Fundação Calouste Gulbenkian, 1999. p. 1272.

[36] Gbenga Oduntan. "The Never Ending Dispute: Legal Theories on the Spatial Demarcation Boundary Plane between Airspace and Outer Space". *Hertfordshire Law Journal*, 1(2). Hertfordshire, England, 2003. p. 70.

[37] Nicolas Mateesco Matte. *Deux Frontières Invisibles: De la Mer Territoriale à l'Air 'Territorial'*. Paris, France: Pedone, 1965.

[38] Gyúla Gál. "Fundamental Links and Conflicts between Legal Rules of Air and Space Flights". *Proceedings of the Twenty-Sixth Colloquium on the Law of Outer Space*. IISL, Budapest, 1983.

[39] A/AC.105/C.2/SR.316 (4.4.79), p. 2.

[40] Jerome Morenoff. *World Peace Through Space Law*. Charlottesville, USA: Michie Co., 1967.

international custom favorable to the functionalist approach was crystallized during the last decades, justifying its application in case of disputes.[41]

Following the functionalist approach, due to its pragmatism, it wouldn't' be proper to make reference to Air Law or Space Law, but to "Aeronautical Law" and "Astronautical Law", as indicated by Jean-Paul Pancracio.[42] For Gyúla Gál, Space Law should be applied as a functionalist system, to rule international relations related to space activities, which, by his understanding, encompass those that have the purpose of allocating objects in orbit around our planet and other celestial bodies, including their launch and return.[43]

Henri A. Wassenbergh stressed that the functional approach excluded the necessity of delimitation of outer space, since space activities would be considered legal as long as conducted pursuant to specific rules, irrespective of where they may take place. Whenever, in accordance with International Law, those activities would implicate in a right of innocent passage for space objects through the air space of other States.[44] In the same sense, Leslie I. Tennen proposed the conventional recognition of the right of innocent passage for every object destined for the outer space, independent of its altitude.[45]

The right of free access to outer space was also mentioned by Nicolas Mateesco Matte, who defended that innocent techniques, which do not cause harm to others in its effort to ignite a object to space, should be considered good, valid and accepted by States. If, during their trajectory, such an object incidentally crossed the air space of other States, it would not be cause for complaint as far as national security is concerned, irrespective of their altitude. On the other hand, whenever a territorial State has enough reasons to suspect that a foreign space object can cause damage to it or is being applied for aggressive means, it would have the right to apply defensive measures proportionate to the threat.[46]

Spatialist advocates frequently critique the vast range of arguments available to States for justification of their actions under the functionalist approach, regardless of the interests of the international community as a whole. Indeed, one of the major obstacles to the general acceptance of the functionalist theory is that, at least initially, it would be up to the State responsible for the activity to determine its nature and, therefore, its legality. Eventual allegations of violation of national territorial integrity, in accordance with the functionalist approach, would demand the recourse to methods of pacific solution of disputes among States, what, according to

---

[41] S. B. Rosenfield. "Where Air Space Ends and Outer Space Begins". *Journal of Space Law*, Vol. 7, n. 148. Mississippi, USA, 1979.

[42] Jean-Paul Pancracio. *Droit International des Espaces*. Paris, France: Armand Colin, 1997. p. 60.

[43] Gyúla Gál. "The Question of Delimitation – After Twenty Years". *Proceedings of the Twenty-Second Colloquium on the Law of Outer Space*. IISL, Munich, 1979. p. 128.

[44] Henri A. Wassenberg. *Principles of Outer Space in Hindsight*. USA: Springer, 1991. p. 18.

[45] Leslie I. Tennen. "Conflicts of Law and Delineation of Outer Space: an Interest Analysis Approach". *Proceedings of the Twenty-Second Colloquium on the Law of Outer Space*. IISL, Munich, 1979.

[46] Nicolas Mateesco Matte. *Aerospace Law*. Toronto, Canada: Carswell, 1969. p. 71.

Bin Cheng, is not Law, but equity, understood in the Aristotelean sense of justice. In the opinion of Cheng, the legal insecurity that the adoption of the functionalist approach would cause should be avoided at all costs, since it disregards the structures of International Law, organized as a system of rules applicable to relations between States that are comprehended as territorially organized entities.[47]

As seen by J. E. S. Fawcett, the functionalist approach has its "dubieties".[48] Francys Lyall and Paul B. Laursen presented three clear objections to the referred theory: first, there could be ambiguity and even dispute among States as to the classification of a particular activity; second, it would not be sensible to omit the location of an activity from consideration of its legality, solely relying on claimed intentions; finally, it should be obtained a solution to the boundary question that can survive future technological developments.[49]

Damodar Wadegaonkar advises that the "letting-sleeping-dogs-lie" tactic defended by functionalists who conceive no need to address the delimitation question right now, may, in fact, turn to be utterly dangerous: "as something will have to be done about it in due course, it would be more prudent to try and slip the muzzle on the animal while all the issues are still dormant than to wait till some major international instance hardens the divergent interests of all nations."[50] As indicated by Nguyen Quoc Dinh, Patrick Daillier and Alain Pellet, the functionalist approach seems to be destined to suffer from the challenge offered by aerospace objects that can operate in air space as well as in outer space.[51]

Among the major functionalist arguments can be found the observation that, until now, a reasonable distance still remains between the maximum altitude that can be reached by aircraft and the minimum perigee of space objects. Commercial aircraft commonly fly through the troposphere, an air layer that, in general terms, extends up to 12 km of altitude in medium latitudes, therefore being considerably lower than the so-called "low Earth orbit", that starts around 90 km above sea level. In view of such fact, the argument goes, any attempt of delimitation of the air space/outer space boundary is unnecessary.[52] But even this functionalist excuse may be put into question, since the maximum altitude that can be achieved by an aircraft is, by a matter of fact, already reasonably close to the lowest orbit conceivable for a space object. It's worth mentioning that the experimental American aircraft X-15

[47] Bin Cheng. *Studies on International Space Law*. Oxford, England: Clarendon Pr, 1998. p. 442/445.

[48] J. E. S. Fawcett. *Outer Space: New Challenges to Law and Policy*. Oxford, England: Clarendon Press, 1984. p. 17.

[49] Francys Lyall and Paul B. Laursen. *Space Law: a Treatise*. Farnham, England: Ashgate, 2009. p. 170.

[50] Damodar Wadegaonkar. *The Orbit of Space Law*. London, England: Stevens & Sons, 1984. p. 47.

[51] Nguyen Quoc Dinh; Alain Pellet; Patrick Daillier. *Direito Internacional Público*. Lisbon, Portugal: Fundação Calouste Gulbenkian, 1999. p. 1272.

[52] To Gbenga Oduntan, that would be another theory, that he calls "the no present need theory". Gbenga Oduntan. "The Never Ending Dispute: Legal Theories on the Spatial Demarcation Boundary Plane between Airspace and Outer Space". *Hertfordshire Law Journal*, 1(2). Hertfordshire, England, 2003. p. 66.

"flew" above 62 miles (100 km) of altitude in 1963, while some satellites have orbited below 69 miles (111 km), as indicated by Glenn H. Reynolds and Robert P. Merges.[53]

It must be stressed that the terms "aeronautical activity" and "space activity", essential for the functionalist approach, were not defined by any treaty. So, if the functionalist approach should prevail, one would only change one lacuna for another: instead of discussing the delimitation of air space and outer space, the debate would be commuted to the necessity of distinguishing aeronautical activities from space activities. When aerospace vehicles are taken into the consideration, the problem reaches even a further level of complexity. Due to their hybrid nature, a definitive distinction of their activities would have to be determined in order to access the legal regime applicable to their operation, a rather difficult challenge considering the multiple tasks they may conduct.[54]

As a matter of fact, a proposal (of treaty or resolution) based on the functionalist approach still remains to be presented before the UNCOPUOS, that presents the legal foundations necessary for its proper adoption, particularly in the case of international disputes regarding the vertical limit of State sovereignty.

On the other hand, several proposals were presented at that international forum to provide a boundary between air space and outer space, based on arguments that range from the scientific rationale to the political and legal ones.[55] According to I. H. Ph. Diedericks-Verschoor, no less than 27 proposals had been submitted by the UNCOPUOS legal subcommittee by the turn of the century.[56] Considering the complexity of some of those suggestions, a deeper consideration of their features is decidedly justified, in order to reckon their pros and cons.

---

[53] "Below an altitude of approximately 69 miles, sustained orbit is practically impossible. Above an altitude of approximately 53-62 miles (the so-called von Karman line) aerodynamic lift is largely nonexistent. Yet X-15 aircraft flew higher than 62 miles, and satellites and other spacecraft pass through orbits lower than 69 miles – and, once experiments with tethered satellites have borne fruits, may stay there for some time". Glenn H. Reynolds and Robert P. Merges. *Outer Space: Problems of Law and Policy*. 2. ed. Boulder, USA: Westview Press, 1997. p. 11/12.

[54] George Paul Sloap. "Relationship of Air Law and Space Law – a View from the Space Shuttle, including its Internal and External Environments". *Proceedings of the Nineteenth Colloquium on the Law of Outer Space*. IISL, New York, 1977. Tanja Masson-Zwaan. "The Aerospace Plane", *In*: Tanja Masson-Zwaan e Pablo Mendes De Leon. *Air and Space Law: De Lege Ferenda – Essays in Honour of Henri A. Wessenbergh*. Dordrecht, the Netherlands: Martinus Nijhoff, 1992.

[55] "Qual será o 'limite vertical' da soberania do Estado? Desde a tese de KROELL, para quem a soberania vai até ao limite da gravidade, à tese de KOVALEV, segundo a qual aquele limite se situa no ponto onde acaba o poder efetivo do Estado, muitas outras surgiram, como a de SACHTER, para quem só haveria soberania até às vinte ou vinte e cinco milhas, ou seja, até onde fosse possível o voo de aviões e à de HINGOMARI, para quem não existe qualquer limite". Albino de Azevedo Soares. *Lições de Direito Internacional Público*. 4. ed. Coimbra, Portugal: Coimbra Editora, 1988. p. 261.

[56] I. H. Ph. Diedericks-Verschoor. *An Introduction to Space Law*. 2. ed. Dordrecht, the Netherlands: Kluwer Academic Publishers, 1999. p. 18/20.

Chapter 4
Proposals

> The path towards the successful delimitation of outer space from territorial space is doubtless narrow and stony (…). [Bin Cheng], Studies in International Space Law[1]

Below are presented the proposals for solution of the long standing debate regarding the vertical limit of State sovereignty, summarized in categories for a better understanding of their main features. Each alternative can be considered as pertaining to the spatialist approach, since they are all based on the premise that the delimitation of the air space/outer space boundary is necessary. The international law scholars that backed them, as well as the States that argued in their favor before the UNCOPUOS, shall be properly referred to in the course fo this chapter. The order of presentation of the suggested proposals follows, as much as possible, a chronological order, although some of the older proposal were many times reassessed years later by different parties, with minor adaptations, to address new political concerns.

The sophistication of some delimitation proposals may impair their full comprehension if only brief remarks are presented; quite a few have their foundations in complex scientific concepts, which will be examined whenever required.

Atmospheric Limit and Aerodynamic Lift

Some law scholars and delegations to the UNCOPUOS have defended a delimitation linked to the maximum altitude which an aircraft is capable of sustaining aerodynamic lift, due to the existence of layers with enough density to support such physical reaction. Thus, the upper limits of our planet's atmosphere would, in one way or another, represent Air Law's jurisdiction, based on the premise that States

[1] Bin Cheng. *Studies on International Space Law*. Oxford, England: Clarendon Pr., 1998. p. 453.

© Springer International Publishing Switzerland 2015

O.O. Bittencourt Neto, *Defining the Limits of Outer Space for Regulatory Purposes*, SpringerBriefs in Space Development,
DOI 10.1007/978-3-319-16685-8_4

have complete and exclusive sovereignty over the column of air above their territories. Beyond the atmosphere, outer space should commence, free from national claims.

During the period that preceded the regular use of Earth's orbit, the atmospheric limit was frequently referred to as a satisfactory value by several UN member States. The UK delegation, in 1959, affirmed that it could be possible to provide a delimitation based on higher atmospheric layers,[2] receiving support by other delegations, including Brazil.[3]

Among law scholars, the Brazilian professor Haroldo Valladão, who, as early as 1958, argued that the realm of Air Law should be connected to the atmospheric space; beyond it, outer space would begin, subjected to a new system of international rules, called by him an "*inter gentes* planetary law".[4] Of similar opinion, Loftus Becker identified the upper atmospheric limit at 10,000 miles of altitude (16,090 km),[5] while Pitman B. Potter advocated that such mark should be delimited at 30 miles high (48.27 km),[6] just to mention two opinions that emphasized the different criteria adopted early on for identification of the atmospheric edge. Michael Aaronson even stated that, since air may residually be found at up to 60,000 miles (96,540 km) above mean sea level, such altitude should be the boundary of territorial sovereignty.[7]

Alex Meyer categorically defended the idea that State sovereignty should extend up to the heights where air can be found.[8] Following his lead, Welf Heinrich, Prince of Hannover, linked the concept of air space to atmosphere, though concluding that a delimitation could only be provided based on concrete factors: the capacity of effective control and the existence of clearly established boundaries.[9] Other important contributions in favor of such a thesis were presented by K. H. Böhme, R. H. Mankiewicsz, Raymond Probst and C. E. S. Horsford; soviet publicists, including D. B. Levin, G. P. Kaluzhnaya and V. I. Lisovsky, also positioned

---

[2] A/AC.98/C.2/SR.1.

[3] A/C.1/SR.986. At the same 1959 session, were favourable to the English position the delegations of Austria, Italy, the Netherlands and Peru.

[4] Haroldo Valladão. "The Law of Interplanetary Space". *Proceedings of the Second Colloquium on the Law of Outer Space*. IISL, London, 1958. p. 157/159.

[5] Loftus Becker. "Major Aspects of the Problem of Outer Space". *Bulletin of the Department of State*. Washington, USA, 1958.

[6] Pitman B. Potter. "International Law of Outer Space". *American Journal of International Law*. Washington, USA, 1958. p. 304.

[7] Michael Aaronson. "Space Law". *International Relations*, vol. I, nº. 9. New York, USA, April/1959. p. 420.

[8] Alex Meyer. "Die Staatshoheit im Luftraum und die Entwicklungen im Weltraum". *ZLRW*, Cologne, Germany, 1965. p. 27.

[9] In another passage of his paper, Welf Heinrich presented support to the "von karman line", position that he would eventually defend throughout his life. "Problems in Establishing a Legal Boundary between Air Space and Outer Space". *Proceedings of the First Colloquium on the Law of Outer Space*. IISL, The Hague, 1957. p. 29.

themselves in support of the reasoning that national sovereignty should be extended up to the places where air could be identified.[10]

In 1963, a group of experts from the *David Davies Memorial Institute of International Studies* presented a project of international code regarding the use of outer space containing provision that defined air space as the volume of space between the surface at sea level and an altitude of 80 km above it, taking into consideration the physical characteristics of air space and outer space and their respective uses.[11]

Then in 1965, C. Wilfred Jenks suggested that the atmospheric limits should represent the limit between air space and outer space, when and if such a distinction proved to be necessary. In his opinion, the best option would be to consider the troposphere and the stratosphere as air space and everything above it as outer space. Nevertheless, Jenks already foresaw the debates that his offer would generate, as the exploration and exploitation of outer space gained momentum:

> A sharp divergence of view concerning both the principle and its application must be expected between those for whom the primary consideration is to secure the greatest possible measure of freedom in space and those for whom the primary consideration is to secure a maximum national control for security or other reasons.[12]

Those proposals rely on a scientific consideration, that is, the measurability of the Earth's atmosphere uttermost limit. But, as alerted by the scientist and writer Arthur C. Clarke, our atmosphere in truth never completely disappears, only dimming with distance, like a musical note fades away with time and space, reaching a level so low that its own existence cannot be sensed.[13]

The reference to scientific studies is considered justifiable, in order to achieve a better understanding of Earth's atmosphere nature and properties.[14] According to Emerson Faria Cabral Paubel, the four most important layers of our atmosphere,

---

[10] Gyúla Gál. *Space Law*. Leiden, the Netherlands: A. W. Sijthoff, 1969. p. 74/75.

[11] In the official commentaries to the proposals, it was clarified: "at the present time the lower effective limit of perigee is in the region of the altitude of 100 miles, since below that the life of the satellite is too short to be useful, and it is possible that an altitude of about 70 miles would be the limit for effective orbiting, since below that friction would become too great. The notion of effectiveness here is to be understood in terms of scientific uses of spacecraft. (…) Any particular altitude chosen as the limit of sovereignty over the airspace may appear arbitrary and be controversial, but, for the avoidance of excessive claims and by the other foregoing considerations, the relatively low altitude of about 50 miles is suggested here as the limit of sovereignty and the beginning of outer space." The David Davies Memorial Institute of International Studies was integrally included as annex X to C. Wilfred Jenks: *Space Law*. New York, USA: Frederick A. Praeger, 1965. p. 419/439.

[12] "C. Wilfred Jenks. *Space Law*. New York, USA: Frederick A. Praeger, 1965. p. 191".

[13] Arthur C. Clarke. *A Exploração do Espaço*. São Paulo, Brazil: Melhoramentos, 1959. p. 26.

[14] In 1976, the Committee on Space Research (COSPAR) of the International Council of Scientific Unions presented study to COPUOS where it was identified the considerable diminishing of atmospheric density between 90 and 150 km of altitude above sea level, from 3.4 mg/m^3 to 0.002 mg/m^3. A/AC.105/164, Annex 1.

that is, troposphere, stratosphere, mesosphere and thermosphere, have specific features, but it cannot be exactly identified where one ends and the other begins.[15]

The gaseous envelope that covers the surface of the Earth, i.e., the atmosphere, as explained by Kshudiram Saha, does not hold a clearly delimited upper limit, and gradually merges into interplanetary space. It is composed of different types of gases, among which are included nitrogen and oxygen, but in order to fully comprehend the thermodynamic properties of the atmosphere, other components must also be taken into account, for instance, water vapor, carbon dioxide and ozone. According to Saha, "the above-mentioned composition of the Earth's atmosphere undergoes changes from above 25 km upward, under the effect of the sun's ultraviolet radiation."[16]

Concurrently, an alternative approach still considering the atmosphere as a relevant factor gained attention, though relying on the vertical limit of aerodynamic lift as the key feature. As a matter of fact, the definition of aircraft provided by the Chicago Convention of 1944, annex 7, amended by ICAO in 1967, clearly establishes a connection with aerodynamic capabilities, as observed by I. H. Ph. Diedericks-Verschoor.[17] In order for an aircraft to fly, the air pressure above its wings shall be lower than the one below, through a physical reaction created by the

[15] "A atmosfera que cobre a Terra é dividida em quatro regiões principais, com diferentes características entre si. Cada camada forma uma espécie de casca esférica que possui uma espessura distinta, mas não se pode estabelecer limites rígidos de onde começa e termina cada uma delas; pelo contrário, as características de cada camada são transferidas para a próxima. Entretanto, as diferenças entre elas são reais. Pela ordem de distância a partir da superfície temos a troposfera, a estratosfera, a mesosfera e, finalmente, a termosfera. À medida que nos afastamos da superfície, a densidade dos gases em cada camada diminui exponencialmente e, acima dos 1.000km de altitude, praticamente impera o vácuo do espaço". Emerson Faria Cabral Paubel. *Propulsão e Controle de Veículos Aeroespaciais*. Florianópolis, Brazil: UFSC, 2002. p. 61.

[16] Kshudiram Saha. *The Earth's Atmosphere: its Physics and Dynamics*. New York, USA: Springer, 2008. p. 10.

[17] "The Paris Convention [1919] contained the first generally accepted definition of the term 'aircraft', which read as follows: 'Le mot aéronef désigne tout appareil pouvant se soutenir dans l'atmosphére grâce aux reactions de l'air'. This rather sweeping definition included aircraft, airships, gliders, free balloons, barrage balloons and helicopters. The criterion which should have been given preference is whether the machine has any lift. Having become outdated, the Convention was eventually replaced, in 1944, by the Chicago Convention, but the latter failed to bring about a change in the definition of 'aircraft', reading: 'Aircraft is any machine that can derive support in the atmosphere from the reactions of the air'. As most other Conventions lacked such a definition altogether the Chicago formula, which had been taken from the Paris Convention, continued to serve as a cornerstone of air law for another few decades, although authoritative opinions were also taken into account on a number of occasions. Eventually, on November 6, 1967, ICAO brought out a new definition reading: 'Aircraft is any machine that can derive support in the atmosphere from the reactions of the air other than the reactions of the air against the earth's surface'. *Its distinctive feature was that the words 'other than the reactions of the air against the earth's surface' had been added. This addition ensured that hovercraft were excluded from the definition of 'aircraft'.*" I. H. Ph. Diederiks-Verschoor. *An Introduction to Air Law*. 5. ed. Dordrecht, the Netherlands: Kluwer Academic Publishers, 1993. p. 5.

shape of such wings. Aerodynamic lift depends upon air density and aircraft speed in order to be properly measured.[18]

Oscar Schachter understood that the air space should be defined in strict connection to the atmospheric elements that enable aerodynamic lift. Hence, where the air is too thin to support the use of an aircraft, for instance, according to the present technological capabilities, from the ionosphere upwards (approximately located around 80 km of altitude above sea level), there should be no State sovereignty.[19] Supporting the same reasoning, Gerhard Reintanz posited, based on technical information, that the gaseous consistency of the atmosphere would end at 100 km of altitude, thus representing the vertical limit of air space.[20]

NASA's scientist Robert Jastrow suggested as early as 1958 that the vertical limit should be established at 100 miles (160.9 km) of altitude, above which there would be outer space, i.e. the region used by space vehicles placed on Earth orbit or beyond.[21] Willy Ley, also taking into consideration scientific data, understood that the altitude of 50 km could provide a reasonable mark to State sovereignty, since up to such level aeronautic activities could be conducted, despite the fact that the planet's atmosphere extends itself up to 250 km of altitude.[22]

In 1962, the Italian delegation at the UN defended a delimitation based on the maximum altitude aircrafts and balloons could fly.[23] Nevertheless, no vertical mark was proposed then, due to the apparent lack of consensus on a specific figure.

Still during the early years of space exploration, Luiz de Gonzaga Bevilacqua, one of the Brazilian space law pioneers, endorsed in international workshops that the exclusive competence of States should be related to the maximum altitude an aircraft could sustain flight, due to physical reactions with the atmosphere.[24]

This position recently regained relevance. Interestingly enough, just a few years ago, such approach was espoused by the UK, in an official document that stated:

> National sovereignty applies to airspace; no sovereignty applies in outer space. There is no established definition of height at which airspace ends and outer space begins. (…) For practical purposes the limit is considered to be as high as any aircraft can fly.[25]

---

[18] Marietta Benkö and Kai-Uwe Schrogl (eds.). *International Space Law in the Making: Current Issues in the UN Committee on the Peaceful Uses of Outer Space*. Paris, France: Frontières, 1993. p. 113/114.

[19] Oscar Schachter. "Legal Aspects of Space Travel". *J. B. I. P. S.*. London, England, 1952. p. 14.

[20] Gerhard Reintanz. "Air Space and Outer Space". *1961 Symposium*, p. 1134.

[21] Robert Jastrow. "Definition of Air Space". *Proceedings of the First Colloquium on the Law of Outer Space*. IISL, the Hague, 1957. p. 82.

[22] Willy Ley. *Rockets, Missiles and Space Travel*. 2. ed. New York, USA: Viking, 1961.

[23] A/C.1/SR.1211.

[24] Luiz de Gonzaga Bevilacqua. "A Contribution to the Problem of Space Law Establishing a Technical and Practical Limit to Political Sovereignty in Space". *Proceedings of the First Colloquium on the Law of Outer Space*. IISL, The Hague, 1957. p. 33.

[25] "National sovereignty applies to airspace; no sovereignty applies in outer space. There is no established definition of height at which airspace ends and outer space begins. (…) For practical purposes the limit is considered to be as high as any aircraft can fly". *UKMIL*, 70 BYIL, 1999, p. 520/521.

It must be stressed that the aerodynamic lift standard retains relevance among publicists. William R. Slomanson, in his International Law handbook published in 2003, sustained that the territorial sovereignty is limited to navigable airspace, represented by the highest altitude that can be achieved by an aircraft.[26] As admitted by the author, the same reasoning was put forth by J. C. Hoogan during the late 1950s, displaying that the foundations for such position rests upon views expressed more than half a century ago.[27]

The scientific support for delimitation based on aerodynamic lift would be arguably be supplied by a study presented by Theodore von Kármán, Hungarian physicist at the University of California, USA, who, in 1957, identified elements for calculation of the specific altitude where aerodynamic pressure is overcame by ascension pressure. The aircraft movement through air space was represented by an elegant equation, where weight = aerodynamic lift + centrifugal speed. Higher from the surface of the Earth, von Kármán realized that the air pressure reduces to a level where only centrifugal force would be applicable, depending upon the speed and trajectory of its object.[28]

Such vertical limit would be later identified as the von Kármán line, often discussed during the early years of space exploration, and especially advocated by Andrew G. Haley, to whom such theoretical formulation represented the best scientific criterion for differentiating air space from outer space. Haley adverted that, in this highly technical field, lawyers could be found in an awkward position, defending thesis without support from basic scientific rules. Therefore, it was suggested by him that, in order to achieve a clear delimitation between air space and outer space, a necessity for identification of the vertical sphere of application of national legal systems, the von Kármán equation emerged as the better alternative to identify such elusive natural limit.[29] In Brazil, C. A. Dunshee de Abranches adopted the von Kármán line in his 1964 study about international liability of States for damages caused by space activities, considering it an useful tool for identification of the different legal regimes envisioned by him for damages caused in air space and outer space.[30]

Based on the von Kármán line elements, Elisabeth Back Impallomeni concluded that, nowadays, International Law recognizes that the limit between air space and outer space is somewhere around 83 km of altitude from sea level, the highest altitude possible for aeronautic flight and the initial position of orbital trajectory.[31]

---

[26] William R. Slomanson. *Fundamental Perspectives on International Law*. 4. ed. Belmont, USA: Wadsworth/Thomson Learning, 2003. p. 286.

[27] J. C. Hogan. "Legal Terminology for the Upper Regions of the Atmosphere and Space Beyond the Atmosphere". *American Journal of International Law,* n. 51. New York, USA, 1957. p. 362.

[28] Robert F. A. Goedhart. *The Never Ending Dispute: Delimitation of Air Space and Outer Space*. Paris, France: Frontières, 1996. p. 61/63.

[29] Andrew G. Haley. *Space Law and Government*. New York, USA: Appleton-Century-Crofts, 1963. p. 97.

[30] C. A. Dunshee de Abranches. *Espaço Exterior e Responsabilidade Internacional*. Rio de Janeiro, Brazil: Livraria Freitas Bastos, 1964. p. 98.

[31] Elisabeth Back Impallomeni. *Spazio Cósmico e Corpi Celesti nell'Ordinamento Internazionale*. Padova, Italy: CEDAM, 1983. p. 88.

Although lying at an average altitude of 80–85 km above mean sea level, the scientific elements for calculation of the von Kármán line would implicate in a variable standard, always subjected to scientific developments. Such reality was even recognized by its main advocate, Andrew G. Haley, who admitted that improvement of fuel refrigeration technics and development of more efficient heat resistant materials could significantly change the location of the von Kármán line.[32]

Scientific experiments designed to measure the boundary between air space and outer space continue to be promoted around the world. A recent paper published in 2009 at the *Journal of Geophysical Research*[33] by Laureline Sangalli and David Knudsen, of the University of Calgary, Canada, presented a possible approach for the distinction of outer space in relation to air space. This is the limit where relatively weak atmospheric winds are overwhelmed by the violent flux of electrically charged particles flowing through outer space, as measured by the satellite JOULE-II. The authors submitted that, in accordance to the data gathered, outer space would begin at approximately 118 km above mean sea level.[34] The research was designed to help scientists to understand the interaction between outer space and our planet's environment, especially in relation to climate change. It could also be useful for the technical delimitation of national sovereignty's vertical frontier, though clearly attesting the difficulties involved in the identification of a physical, permanent and clear boundary.[35]

In one opinion, the constant evolution of technology would imply the benefit of having a fluid boundary, if based solely on the atmospheric or aerodynamic lift limits.[36] Studies show that, even though our planet's atmosphere is concentrated around 90–100 km above the ground, residual air may be found at considerably high

---

[32] "The von Karman primary jurisdictional line may eventually remain as presented above [275.000 feet] or, as a result of such developments as improved techniques of cooling and more heat-resistant materials, it may be significantly changed. But these changes will be only in the exact location of the von Karman line, for the existence of the line is certain, and wherever it is finally drawn will be the place where 'airspace' terminates." Andrew G. Haley. *Space Law and Government.* New York, USA: Appleton-Century-Crofts, 1963. p. 98/99. In the same way: "Improved techniques of cooling and more heat-resistant materials may eventually change the exact location of the von Kármán line; the existence of the line itself is undisputed." Robert F. A. Goedhart. *The Never Ending Dispute: Delimitation of Air Space and Outer Space.* Paris, France: Frontières, 1996. p. 63.

[33] Journal of Geophysical Research. http://www.agu.org/pubs/crossref/2009/2008JA013757.shtml. Published on 04.07.2009. Accessed on 04.20.2010.

[34] Laureline Sangalli and David Knudsen. "Rocket-based Measurements of Ion Velocity, Neutral Wind, and Electric Field in the Collisional Transition Region of the Auroral Ionosphere". *Journal of Geophysical Research,* 2009; 114 (a4): A04306.

[35] University of Calgary. "Scientists Pinpoint the 'Edge of Space'". *ScienceDaily.* Published on 04.09.2009. http://www.sciencedaily.com/releases/2009/04/090409142301.htm, accessed on 04.20.2010.

[36] As a historic reference, one can mention the opinion presented by Marcel Le Goff, pioneer of Air Law, who, before the space age, presented the "biological theory", stating that national air space would reach only the maximum altitude where a human being can breath. *Traité Théorique et Pratique de Droit Aérien.* Paris, France: 1958. p. 20. In the same senseo: Hubert Strughold. "Definitions and Subdivisions of Space: Bioastronautical Aspect". *Proceedings of the First Colloquium on the Law of Outer Space.* IISL, The Hague, 1957.

altitudes. Besides, the atmosphere suffers a cyclical movement similar to the tide effect, due to the gravitational pull from nearby celestial bodies, specially the Moon, and also from shifting energy levels from the Sun.[37] With proper alterations, objects could eventually produce aerodynamic lift with very little air resistance at extreme altitudes. Furthermore, it must be acknowledged that establishing a delimitation based on criteria related to the extension of Earth's atmosphere would, sooner or later, present challenges to space activities, since some of the most important orbits used by space objects traverse our planets exosphere, hundreds of kilometers above the ground.[38]

Lowest Orbital Perigee

Following in an opposite direction are the proposals defending delimitation of the outer space at the lowest perigee achievable by space objects, that is, the point of their orbits where they are closest to the Earth's surface.[39]

Since there is agreement among States that an object in orbit does not disrespect national territorial integrity, in fact an established international custom at least with regard to the maiden Sputnik mission, in 1957,[40] this position would be supported by general practice. Indeed, the Soviet position ever since the success of the first soviet space excursions, as stated by G. P. Zadorozhny, claims that space objects do not violate the air space of other States, since their territories are not overflown, but only transversed by successive orbital trajectories dictated by the rotation of the planet and connected to the inner speed of the artificial satellites.[41]

During the months following the "dawn of space age", the delegation of Peru stated at the UN that, since no State had protested the invasion of their territories by Soviet space objects, Earth orbit could not be subjected to national claims.[42]

---

[37] As explained by the scientists Camil Gemael and José Bittencourt de Andrade, the air density is an important coefficient for calculation of drag over satellites, which affect their orbits: "hoje sabemos que essa densidade varia em função da atividade solar, por vezes de maneira súbita e rápida, o que traz complicações ao problema quando se buscam soluções rigorosas." *Geodésia Celeste*. Curitiba: Ed. UFPR, 2004. p. 99.

[38] Bin Cheng. *Studies on International Space Law*. Oxford, England: Clarendon Pr, 1998. p. 6/9.

[39] "The extreme end-points of the major axis of an orbit are referred as 'apses'. The point nearest the prime focus is called 'periapsis' (meaning the 'near apse') and the point farthest from the prime focus is called 'apoapsis' (meaning the 'far apse'). Depending on what is the central attracting body in an orbital situation these points may also be called 'perigee' or 'apogee', 'perihelion' and 'aphelion', 'periselenium' e 'aposelenium', etc." Roger R. Bate, Donald D. Mueller and Jerry E. White. *Fundamentals of Astrodynamics*. New York, USA: Dover, 1971. p. 24.

[40] Peter Malanczuk. *Akehurst's Modern Introduction to International Law*. 7. ed. New York, USA: Routledge, 2006. p. 206.

[41] G. P. Zadorozhny. "Iskustvennye Sputniki i Mezhdunarodnoye Pravo". *Sovietskaya Rossiya*, Moscow, Russia, October/1957. p. 1/2.

[42] A/C.1/SR.983, 13.11.1958. The Swedish delegation presented itself radically against such position: A/C.1/SR.984.

The principle of lowest perigee as the air space/outer space boundary was espoused by D. Goedhius during the International Law Association (ILA) meeting held in Helsinki, 1966, serving as the basis for a resolution approved by said organization at their next meeting, in Buenos Aires, 1968. According to Goedhius, the International Law Association agreed that the States' practice was consistent with the view that sovereignty could not be extended above the lowest perigee of any satellite placed in orbit.[43] Several scholars defended the same reasoning,[44] including Lord Arnold Duncan MnNair[45] and William J. Hughes.[46]

Vladimir Kopal has, for a considerable time, asseverated that the Earth orbit constitutes the best criterion for delimitation of outer space's inner boundary.[47] In his view, despite no formal agreement being yet achieved among States in relation to the definition of the term "outer space", it would have been internationally accepted, expressly or tacitly, that every object in orbit is, as a matter of fact, in outer space.[48]

The lowest perigee appears as the most reasonable standard in a study presented by L. Perek to the IISL in 1976, since it is based on a strong scientific fact, making it capable of prevailing over any other alternative. Later in his paper, the author identified this limit at 100 km of altitude above sea level, measured perpendicularly in relation to the Earth.[49]

In South America, this opinion was defended by the Uruguayan publicist Araújo Bauza, who indicated as the ultimate altitude for application of Air Law and Space Law the lowest orbit possible for artificial satellites, a region he erroneously identified as the limit of Earth's gravity.[50] Reflecting a similar understanding, though more accurate as far as the science is concerned, are the opinions of the Argentinian law scholar Pasini Costadoat.[51]

An interesting approach has been introduced by Russia and China, supporting a restrictive approach to the concept of "space objects", which would be restricted to those that complete at least one orbit around the Earth. In accordance with the

[43] *Report of the 52nd Conference.* ILA, Helsinki, 1966. p. 160/185, 191/201.

[44] Gbenga Oduntan. "The Never Ending Dispute: Legal Theories on the Spatial Demarcation Boundary Plane between Airspace and Outer Space". *Hertfordshire Law Journal,* 1(2). Hertfordshire, England, 2003. 64/84.

[45] Arnold Duncan McNair. *The Law of the Air.* 3. ed. London, England: 1964. p. 15.

[46] William J. Hughes. "Aerial Intrusions By Civil Airliner And The Use of Force". *Journal of Air Law and Commerce.* Dallas, USA, 1980. p. 595.

[47] Vladimir Kopal. "What is 'Outer Space' in Astronautics and Space Law". *Proceedings of the Tenth Colloquium on the Law of Outer Space.* IISL, Belgrade, 1967. p. 277.

[48] Vladimir Kopal. "Issues Involved in Defining Outer Space, Space Object and Space Debris". *Proceedings of the Thirty-Forth Colloquium on the Law of Outer Space.* IISL, Montreal, 1991. p. 40.

[49] L. Perek. "Remarks on Scientific Criteria for the Definition of Outer Space". *Proceedings of the Nineteenth Colloquium on the Law of Outer Space.* IISL, Anaheim, 1976. p. 194.

[50] Araújo Bauza. *Hacia un Derecho Astronáutico.* Montevideu, Uruguay: Roque Desalma, 1957. p. 125.

[51] Pasini Costadoat. *El Espacio Aéreo.* Buenos Aires, Argentina: Depalma, 1955. p. 132.

excellent study on the definition and delimitation of outer space prepared by Marietta Benkö and Engelbert Plescher, such interpretation would therefore exclude the applicability of space law to objects performing suborbital flights, "since all those [suborbital] objects are launched into outer space and make use of the properties of the outer space environment for their flight and for reaching their destination, there does not seem to be a valid argument against their qualification as 'space objects' in the framework of international law."[52]

One should bear in mind that, as explained by J. E. S. Fawcett, the fact that a satellite in orbit is accepted as being beyond the sovereign territory of States does not necessarily lead to the conclusion that the actual border of outer space is solely connected to the lowest orbital position in relation to the Earth:

> (…) two propositions can hardly be disputed: that a principle of free and common use precludes the exclusive jurisdiction of any State; and that a spacecraft in Earth orbit is already in outer space. (…) However, the conclusion that satellites in Earth orbit are in outer space implies not that the lower limit of outer space is necessarily identical with the lowest effective orbiting altitude, but only that it is not higher.[53]

Once again, the problem lies in providing the vertical limit of State sovereignty based on a standard permanently subjected to the human capacity to overcome technical limits as far as the orbit positions are concerned. It also has to be acknowledged that space objects in low Earth orbit, that is, below 5,000 km of altitude from sea level, suffer constant perturbations during their trajectories, mostly due to air drag produced by external layers of our planet's atmosphere.[54] Therefore, the lowest perigee of an artificial satellite will ultimately depend on the capacity of orbital correction, by mathematic calculation and mechanical intervention compensating such disturbances to the object trajectory.

That is the reason why connecting the "final frontier" to the highest altitude attainable by an aircraft or the lowest perigee possible for a space object would not solve the problem, instead only creating an undetermined and fluid boundary, permanently under threat of further revision.

Gravitational Limit

Other spatialists suggested that outer space should commence where Earth's gravitational field cease to be effective. Therefore, some words should be spent considering the nature of gravity. Such a force was identified by Sir Isaac Newton

---

[52] Marietta Benkö and Engelbert Plescher. *Space Law – Reconsidering the Definition/Delimitation Question and the Passage of Spacecraft Through Foreign Airspace.* The Hague, The Netherlands: Eleven, 2013. p. 37.

[53] J. E. S. Fawcett. *International Law and the Uses of Outer Space.* Manchester, England: University Press, 1969. p. 17 and 18.

[54] Roger R. Bate, Donald D. Mueller and Jerry E. White. *Fundamentals of Astrodynamics.* New York, USA: Dover, 1971. p. 152/159.

(1642–1727) who, standing on the shoulder of the great scientists that preceded him, revealed the laws of universal gravitation in his magnum opus "Mathematical Principles of Natural Philosophy" (1687), universally recognized by the first word of its Latin title, *Principia*.[55] The contribution of the titan Newton to science is incommensurable, according to Antônio Manuel Alves Morais, who, in a scientific study, explained how the British genius established that the gravitational effect applies to every celestial body, even the Sun, which cannot be kept in the center of the Solar System, since it is under the forces of other celestial bodies.[56]

The law of gravity, as identified by Newton, obeys an elegant mathematic formula: every particle in the Universe attracts all other particles with a force that is directly proportional to the product of the masses of those particles and inversely proportional to the square of the distance between them.[57] Since representing an universal rule applicable to every celestial body, it explains the tide effect, happening due to gravitational disturbance created by the Sun and especially the Moon on Earth's oceans and seas.[58]

Later, Newton's lessons would be revised by another magnificent virtuoso of science, Albert Einstein (1879–1955), who, in his General Theory of Relativity, indeed a theory on gravity, identified that the apparent "attraction" among celestial bodies actually happens due to the warping of space and time by the more massive one in relation to the others. One must recall that Einstein also recognized the intrinsic relationship between mass and energy (according to the famous equation $E=mc^2$).[59]

---

[55] "No desencadeamento da teoria da gravitação, ocorreu um encadeamento de trabalhos que culminarão na obra monumental de Isaac Newton: 'Princípios matemáticos de filosofia natural', conhecido vulgarmente pelo seu primeiro nome em latim, Principia. São importantes as discussões feitas por Christopher Wren (1632-1723), Edmond Halley (1656-1742) e Robert Hooke (1635-1702) sobre as leis que governavam as formas precisas das órbitas dos objetos celestes e as suas relações com as leis da mecânica, tais como a lei de inércia e a lei de força central e atrativa (dirigida para o Sol)." Antônio Manuel Alves Morais. *Gravitação & Cosmologia: uma Introdução.* São Paulo, Brazil: Editora Livraria da Física, 2010. p. 66.

[56] "[Isaac Newton] estabeleceu que a atração gravitacional atua entre todos os corpos do universo e concluiu que o Sol não pode estar em repouso no centro do universo, pois está sujeito às forças dos outros corpos celestes. Determinou as perturbações nas órbitas planetárias devido aos outros planetas, mostrou que a órbita dos cometas não é irregular, estudou a atração gravitacional de um corpo extenso não esférico (elipsóide de revolução), estabeleceu que a Terra deveria ser achatada e determinou esse achatamento, prevendo a variação gravitacional com a latitude. Propôs um método para determinar experimentalmente esse efeito utilizando pêndulo, e explicou ainda a precessão dos equinócios e as marés." Antônio Manuel Alves Morais. *Gravitação & Cosmologia: uma Introdução.* São Paulo, Brazil: Editora Livraria da Física, 2010. p. 72.

[57] Isaac Newton. *Principia – Princípios Matemáticos de Filosofia Natural (Livro III).* São Paulo, Brazil: Folha de São Paulo, Brazil, 2010. p. 20.

[58] "A força que mantém os corpos celestes em suas órbitas tem sido chamada até aqui de força centrípeta, mas tendo ficado evidente que ela não pode ser outra que não uma força gravitacional, vamos chamá-la daqui por diante de gravidade. Pois a causa desta força centrípeta que mantém a lua em sua órbita estende-se a todos os planetas." Isaac Newton. *Principia – Princípios Matemáticos de Filosofia Natural (Livro III).* São Paulo, Brazil: Folha de São Paulo, Brazil, 2010. p. 20.

[59] Where "E" means energy; "m", mass; and "c", the universal constant, that is, the speed of light in vacuum (299,792,458 m/s).

Consequently, no direct connection or bond between the Earth and space objects in its orbit can be observed; such artifacts obey their path due to the warp produced by our planet to the universal fabric that it is located and to what it belongs.[60]

The explanations provided by the General Theory of Relativity to the concept of gravity prompted a revolution on astrophysics. Additionally, one must realize the consequences to Space Law of Einstein's findings as far as the relationship between bodies in outer space is concerned: no State may claim, based on scientific grounds, that it is the attraction provided by its own territory that keeps a space object in a certain trajectory, even regarding the geostationary orbit (located at a particular distance that an object therein moves around the Earth at the same speed that the planet rotates around itself, in such a way that the satellite appears as remaining still when observed from the ground).

This is because there is no physical bond between any specific segment of the Earth's surface and a space object in orbit. In fact, it is the trajectory of the object that is altered by the whole planet, warping the space and time around it, due to its total mass. The closer to the Earth, the stronger is the distortion—which does not mean that general gravitational effects cannot be observed at substantial distances. Unfortunately, those scientific particularities often were not fully grasped by legislators.

For instance, according to Joseph Kroell, it would not be possible to provide a maximum limit of altitude to sovereignty, which could only be limited by its own capacity of being exercised. In conclusion, "the only frontier recognized by States, vertically wise, is the one that the nature imposes to mankind, that is, the physical limit beyond which no life form is possible. In practice, sovereignty ends in that absolute limit."[61] Later, Kroell proved to be a strong advocate of the gravitational effects' limit, when in 1953 he published his paper "Éléments Créatures d'un Droit Astronautique", where it is stated that only the space beyond the Earth's gravity could be opened to all States.[62]

In Brazil, José Dalmo Fairbanks Belfort de Mattos presented a similar proposal during the II Regional Conference of Air Navigation held by ICAO in the city of São Paulo, Brazil, 1956. As he explained his idea: "State sovereignty shall be exercised in plenitude over the air space, that covers the terrestrial and maritime territories, as well as in relation to interplanetary space, which extends itself up to a 'neutral point' or 'neutral zone'; the over flight, by States' objects, whether guided or not, shall only be allowed: a) by previous and unequivocal authorization by the State overflown; b) by specific international agreement, registered at ICAO Secretary

[60] Albert Einstein. *Relativity: the Special and the General Theory.* New York, USA: Dover, 2010.

[61] "La seule frontière que les États connaissent, dans le sens vertical est celle que la nature impose à l'homme, c'est-à-dire, une limite physique ou delá de laquelle, toute manifestation de vie est impossible. Pratiquement, la souveraineté de l'État se termine avec cette limite absolue". Joseph Kroell. *Traité de Droit International Public Aérien.* Tomo I. Paris, France: Les Éditions Internationales, 1934. p. 81.

[62] Joseph Kroell. "Éléments Créatures d'un Droit Astronautique". *Revue Générale de l'Air*, n. 16. Paris, France, 1953. p. 230 e ss.

and UN General Secretary."[63] Belfort de Mattos clarified that, as far as his proposal is concerned, the term "neutral point" constitutes the place in outer space where the attraction of the Moon would supersede the Earth's gravity, while the "neutral zone" would be the outer space region where Earth's attraction would be practically nulled due to the distance to the ground. "In both cases" concluded Belfort de Mattos, "the interplanetary objects could not 'fall back' into Earth, except in case of controlled return."[64] Recently, another celebrated Brazilian publicist, the late Guido Fernando Silva Soares, supported, in general terms, the same reasoning:

> There are no legal definition of States' airspaces, neither of its external limits, but a criterion for delimitation has been the space where Earth's gravity can be verified and where civil and military aviation as well as ballistic activities take place (beyond which is the outer space (…)).[65]

In an interesting variation of this view, Gerd Rinck understood that the Earth's gravity and, therefore, sovereignty, should be extended up to 1.5 million km from the surface of the planet, though recognizing that, in accordance to such position, most of the possible orbits would become subjected to national jurisdiction. As a consequence thereof, no State would be obliged to accept that a space object used a trajectory above its territory. To the author, nevertheless, the gravitation effects' limit could provide security for States, and would better reflect the concept of sovereignty. Rinck's proposal therefore involves a vertical limit so vast that it would even include the Moon under the sovereign sphere of States, even though he argued that States should contemplate a treaty preventing such appropriation.[66]

This complication to the gravitational effects' proposal for delimitation of outer space would create an unsatisfactory situation, since consequently authorizing national claims over terrestrial orbits is a flagrant breach of a fundamental principle of Space Law.[67]

---

[63] "Que a soberania do Estado se exerça, em sua plenitude, sobre o espaço aéreo, que recobre seus territórios terrestre ou marítimo, bem como sobre o espaço interplanetário, que se estende entre os aludidos limites, até o 'PONTO NEUTRO' ou 'ZONA NEUTRA'; que o sobrevoo, do Estado, por teleguiados ou não de outra potência só será lícito: a) mediante autorização prévia e inequívoca do Estado a ser sobrevoado; b) ou mediante Acordo Internacional específico, registrado na Secretaria da OACI e na Secretaria Geral da ONU." José Dalmo Fairbanks Belfort de Mattos. *Direito Internacional.* São Paulo, Brazil: Saraiva, 1979. p. 116.

[64] "Em ambos os casos, os engenhos interplanetários não mais poderão 'cair' para a terra, salvo em torna-viagem." José Dalmo Fairbanks Belfort de Mattos. *Direito Internacional.* São Paulo, Brazil: Saraiva, 1979. p. 123.

[65] "Inexistem definições jurídicas do que sejam os espaços aéreos dos Estados, quais seus limites exteriores, mas um critério para delimitá-lo tem sido o espaço onde a força da gravidade da terra se verifica e onde se exerça a aviação civil e militar e as atividades militares relacionadas a balísticos (além dos quais se encontra o espaço exterior, também denominado cósmico ou sideral) (…)." Guido Fernando Silva Soares. *Curso de Direito Internacional Público.* 2. ed. São Paulo, Brazil: Atlas, 2004. p. 295.

[66] Gerd Rinck. "Recht im Weltraum". *ZLRW*, Cologne, Germany, 1960. p. 191-208.

[67] es. Celso D. de Albuquerque Mello. *Curso de Direito Internacional Público.* 14. ed. São Paulo, Brazil: Renovar, 2002. p. 1324.

In practice, the gravitational effects from the Earth influence celestial bodies much more distant than the Moon itself. Thomas Gangale remarked the case of the asteroid 3753 Cruithne, with 5 km of diameter, whose orbit, discovered in 1986, comprises a complex trajectory around the Earth and the Sun, even justifying it being occasionally referred as our planet's second moon, despite generally being found thousands of kilometers away.[68]

The greatest fault of the gravitational effects position is the fact that there is no "gravisphere", i.e., no physical limit of Earth's gravitation, as perfectly pointed out by Gyúla GÁL; indeed, in accordance to the laws of nature, it is not possible to determine up the vertical limits of terrestrial "gravity", understanding it as the warping of the fabric of space and time provided by our planet's mass. Therefore, the greatest obstacle to the gravitational effects delimitation is not presented by Law, but by science itself.[69]

Effective Control

Due to the difficult to resort to a clear scientific parameter for delimitation of the vertical limit to State sovereignty, the political and pragmatic standard of effective control has been suggested, as presented by Hans Kelsen.[70] In a similar way, in his classic book "Direito Público Internacional", Clóvis Beviláqua, one of the most relevant Brazilian legal scholars, affirmed that the air space "corresponds to the territory up to the altitude determined by a country's safety and security necessities".[71]

That is the result of applying the famous Cornelius Von Bynkershoek thesis for identifying the territorial sea's limits to Air Law and Space Law of: "dominium terrae finitur ubi finitur armorum vis". In 1898, Emanuel von Ullmann applied it to a then nascent group of legal rules pertaining to aeronautical activities, on the understanding that the power of the territorial State should be extended to all activities it undertakes.[72]

One of the first publicists to address the principle of effective control in the space era, H. B. Jacobini, based his considerations on historical research in order to justify his conclusion that such principle constituted the one best suited for distinguishing the vertical limit of State sovereignty.[73]

---

[68] Thomas Gangale. *The Development of Outer Space: Sovereignty and Property Rights in International Space Law*. Santa Barbara, USA: Praeger, 2009. p. 11.

[69] Gyula Gál. *Space Law*. Leiden, the Netherlands: A. W. Sijthoff, 1969. p. 72.

[70] Hans Kelsen. *Teoria Geral do Direito e do Estado*. São Paulo, Brazil: Martins Fontes, 2005. p. 312.

[71] "O espaço aéreo corresponde ao território, até a altura determinada pelas necessidades da polícia e segurança do país." Clóvis Beviláqua. *Direito Público Internacional*. Tomo I. Rio de Janeiro, Brazil: Livraria Francisco Alves, 1911. p. 277.

[72] Emanuel von Ullmann. *Völkerrecht*. Tubingen, Germany: J. C. B. Mohr, 1898. p. 180.

[73] H. B. Jacobini. "Effective Control as Related to the Extension of Sovereignty in Space". *Journal of Public Law*, n. 7, 1958. p. 97 e ss.

Alfred Verdross, on his part, indicated that since the sphere of State jurisdiction could not be extended beyond its effective power, in accordance with the principle of effective control, a vertical limit to State sovereignty ought to exist. Nevertheless such limit's possible reassessment due to the development of technology, it should not be extended beyond the column of air above the surface of our planet.[74]

Luis Ivani de Amorim Araújo defended this position based on the following reasoning: "State sovereignty extends itself up to the altitude where mankind can, directly or indirectly, reach. Up to the place where a man can go, there is the sovereignty of State, since science shall provide the means to make effective and real the sovereign control."[75] According to the author, his opinions are similar to the ones of Ming Min Peng, who supported the application of such reasoning to aviation, irrespective of the maximum altitude achievable.[76] National legislation occasionally adopted similar provision, for instance the Bolivian Aeronautical Code of October 24th, 1930, which provided that the limits of national airspace would be the maximum reachable by national military forces.

However, it has to be acknowledged that the effective control alternative is beneficial to the great powers, which hold better overall resources to exercise their sovereignty. Therefore, proposals under such reasoning shall not be accepted, due to the direct offense they present to the principle of equality of States, as contemplated by the UN Charter.[77] Indeed, the San Francisco Treaty, in its article 1, paragraph 2, would be breached by the effective control thesis, since, as indicated by W. W. C. de Vries, it would not respect the equality of nations.[78]

If applied as indicated by A. S. Piradov, such a thesis would result in as many vertical limits as States in the world, each one determined by their respective levels of economic, scientific and technological progress.[79] The legal insecurity that it would create cannot be considered beneficial to the development of space activities. Further, a clear offence would be imposed to the interests of less developed States,

[74] Alfred Verdross. *Volkerrecht*. 5. ed. Vienna, Austria: Springer Verlag, 1964. p. 274.

[75] Luis Ivani de Amorim Araújo. *Curso de Direito Internacional Público*. 6. ed. Rio de Janeiro, Brazil: Forense, 1988. p. 207/208.

[76] Ming Min Peng. "Le Vol à Grande Altitude et l'Article 1er de la Convention de Chicago". *Révue Franceise de Droit Aérien*. Paris, France, 1952, n°. 4.

[77] I. H. Ph. Diederiks-Verschoor. *An Introduction to Space Law*. 2. ed. Dordrecht, the Netherlands: Kluwer Academic Publishers, 1999. p. 19.

[78] W. W. C. de Vries. "Boundaries in Space? Some Theories and Problems Regarding the Definition, Delimitation and/or Demarcation of Outer Space". *Thesauros Acroasium*. Institution of International Public Law and International Relations of Thessaloniki, Vol. XIV. Thessaloniki, Greece, 1985. p. 779.

[79] A. S. Piradov. *International Space Law*. Honolulu, USA: University Press of the Pacific, 2000. p. 33.

since for the superpowers, as asserted by Myres S. McDougal, Harold D. Lasswell and Ivan I. Vlasic, "not even the sky would be the limit".[80]

To Nicolas Mateesco Matte one of the major shortcomings of the effective control thesis is that it would "result in the abnormal situation where some States would have higher columns of air (or space) under their jurisdiction than other states, depending on the technical perfection of their weapons. Moreover, the boundary between air and the rest of space would be continually changing as a result of technical developments."[81]

If the effective control were to be applied, higher air spaces would have to be recognized at least for China and USA, which, in 2007 and 2009, performed successful tests of anti-satellite weapons, used for the destruction of space objects beyond 800 km of altitude and producing a considerable amount of space debris in the process.[82] A "Pandora's Box" would be opened, authorizing other States to plan identical tests in order to constantly extend their sovereignty. This would begin an arms race that disrespects the principles of the Space Treaty of 1967 regarding peaceful exploration and use of outer space for the benefit of all mankind.[83]

Arbitrary Delimitation

Finally, some spatialists defend the provision of a conventionally accorded altitude above which outer space would commence. Throughout the years that immediately preceded and followed the inaugural flight of Sputnik I (1957), several proposals were presented for arbitrary vertical demarcation, in a diverse and, many times, conflicting collection, superbly compiled by Gyúla Gál as early as 1969.[84]

---

[80] "The application of such a doctrine with respect to any problem of legal order in the contemporary world community would no doubt be highly dangerous; it would be certainly disastrous in the domain of space. If every state were allowed to project its sovereignty upward and sideward in accordance with its effective Power, there would inevitably arise countless conflicting claims with no criteria for their accommodation other than naked power. Moreover, for many underdeveloped states sovereignty would end at the treetops, while for a handful of the most powerful states, not even the sky would be the limit". Myres S. McDougal, Harold D. Lasswell and Ivan I. Vlasic. *Law and Public Order in Space*. New Haven, USA: Yale University Press, 1963. p. 342.

[81] Nicolas Mateesco Matte. *Aerospace Law*. Toronto, Canada: Carswell, 1969. p. 33.

[82] José Monserrat Filho. *Direito e Política na Era Espacial*. Rio de Janeiro, Brazil: Vieira & Lent, 2007. p. 93/94.

[83] The Mauritius delegation, in its official answer to the questionary prepared by the UNCOPUOS legal subcommittee regarding national position about definition and delimitation of outer space, presented in 2010, affirmed that the country exercises control over aeronautical activities up to 46,000 feet of altitude (14.2 km): "Mauritius does not carry out any outer space activities. With regard to civil aviation activities, airspace up to the level of 46,000 feet in the Flight Information Region is controlled by the Area Control Centre, while the airspace above 46,000 feet remains uncontrolled." A/AC.105/889/Add.8.

[84] The Hungarian celebrated publicists collected proposals from 59 authors, generally with different standards, from 12 km (Beresford) to 1,500,000 km (Rinck). The record is comprehensive and deserves a special reference. Gyúla Gál. *Space Law*. Leiden, The Netherlands: A. W. Sijthoff, 1969. p. 114/116.

At the UN, by that time, many delegations enthusiastically supported the necessity of providing the definition of an altitude limit for the identification of outer space. It shall be remarked the opinions presented by Sweden,[85] Spain[86] and Chile,[87] still in 1958. Argentina argued, in 1961, that the best alternative would be the provision of an arbitrary standard mutually accorded by States, considering, *inter alia*, national security concerns.[88]

The French delegation to the UNCOPUOS, in 1966, concluded a working paper according to which the best criteria to distinguish air space from outer space would be the settling of a certain altitude as ultimate boundary, proposed, at that opportunity, as of 80 km from sea level.[89] By its turn, Canada stated that the maximum altitude of State sovereignty should be either equal to 1/100 of the distance from the Equator line to the poles, i.e., 100 km from sea level, or correspondent to 1/100 of the Earth's ratio, that is, 64 km of altitude.[90] At the same session, Italy and Iran supported delimitation based on 100 km of altitude.[91]

In 1969, the Italian delegation emphasized the necessity of elaboration of a treaty that provided delimitation at 120–150 km of altitude, in order to guaranty a safety margin for eventual navigation errors without causing international controversy.[92] The Austrian delegation, at the same session, presented an opinion favorable to delimitation at 80 km of altitude.[93] Some years later, in 1975, Italy presented to the UNCOPUOS an official proposal suggesting a vertical limit at 90 km of altitude above sea level.[94] At the next session, the Belgian delegation offered an interesting study about "Natural Frontiers in Space", where, based on several scientific data, a suggestion favorable to a 100 km of altitude boundary was presented.[95] Iran stated in 1978 that the technical knowledge of the time allowed delimitation in some point between 80 and 100 km above sea level.[96]

It was 1979 when the USSR presented its official position in favor of an international agreement establishing that, above 100–110 km of altitude, outer space would begin, but including the right of innocent passage below such mark, in relation to launchings of space objects and their return to the ground.[97] Later, in 1983, the proposal was reintroduced by the Soviet delegation, with minor modifications:

---

[85] A/C.1/PV.984 and A/C1/SR.1079.

[86] A/C.1/PV.992.

[87] A/C.1/PV.982.

[88] A/C.1/SR.1211.

[89] A/AC.105/C.1/WP.V.1.

[90] A/AC.105/C.1/SR.44.

[91] A/AC.105/C.1/SR.45.

[92] A/AC.105/C.2/SR.113.

[93] A/AC.105/C.2/SR.114.

[94] A/10020, paragraph 27. Before, in 1958, the Italian delegation to the United Nations had supported the 100 km vertical limit: A/C.1/PV.982. In 1977, Italy explained that its 1975 proposal was flexible, and could be raised up to 100 km: A/AC.105/C.2/SR.269.

[95] "Natural Boundaries in Space". A/AC.105/C.1/L.13.

[96] A/AC.105/C.2/SR.298.

[97] A/AC.105/C.2/L.121. The proposal received support from Bulgaria, Belgium and Chile.

Approach to the delimitation of air space and outer space.

1. The boundary between outer space and air space shall be established by agreement among states at an altitude not exceeding 110 km above sea level, and shall be legally confirmed by the conclusion of an international legal instrument of a binding character.
2. This instrument shall also specify that a space object of any State shall retain the right of peaceful passage over the territory of other. States at altitudes lower than the agreed boundary for the purpose of reaching orbit or returning to earth.[98]

The major change was that the term "innocent passage", included in the 1979 text, was replaced for "peaceful passage", explained by the Soviet delegation as one that does not cause adverse effects to the territory being crossed.[99] After the end of the USSR, the Russian delegation maintained a position favorable to delimitation following the same orientation presented in 1983, as asserted by Yuri Kolossov and Dmitry V. Gouchar. According to them, Russia will continue to defend such a proposal until a binding legal international instrument rules on the topic for the benefit of all mankind.[100]

It is worth mentioning that the renowned scholar Epitácio Pessoa, as early as 1911, when preparing his "Project of International Law Code" ("Projeto de Código de Direito Internacional Público"), included, among its 721 articles, one that specifically dealt with the vertical limit of national air space (article 57): "each State has, over the air space correspondent to its territory, up to 1,500 meters, the rights necessary for its conservation. The zone hereby delimited is call protective zone".[101] In a further provision, included in the draft article 257, Pessoa stated: "States have the right to rule, for the benefit of their security or customs and sanitary concerns, the passage of foreign aircraft through their protective zone, or even to partly forbid for their conservation or defense."[102] As observed by Luis Ivani de Amorim Araújo, the reading of those articles allows one to conclude that "the illustrious Brazilian publicist was of the opinion that State sovereignty, vertically wise, had a limit, above which the space would be completely free".[103]

---

[98] A/AC.105/C.2/L.139.

[99] A/AC.105/C.2/SR.392.

[100] Yuri Kolossov and Dmitry V. Gouchar. "Delimitation of Airspace and Outer Space: a Legal View". *Revista Brasileira de Direito Aeronáutico e Espacial - SBDA*, n. 89. Rio de Janeiro, Brazil, mar.-2006. p. 19.

[101] "Cada Estado tem sobre o espaço aéreo correspondente ao seu território, até a altura de 1.500 metros, os direitos necessários à sua conservação. A zona assim limitada tem o nome de zona de proteção." Epitácio Pessoa. *Projeto de Código de Direito Internacional Público*. Rio de Janeiro, Brazil: Imprensa Nacional, 1911.

[102] "Os Estados têm o direito de regulamentar, em bem da sua segurança ou dos seus interesses aduaneiros e sanitários, a passagem de aeróstatos estrangeiros pela sua zona de proteção, ou mesmo proibi-la nos trechos dessa zona em que o exigirem as necessidades da sua conservação e defesa." Epitácio Pessoa. *Projeto de Código de Direito Internacional Público*. Rio de Janeiro, Brazil: Imprensa Nacional, 1911.

[103] "(...) o ilustre internacionalista brasileiro era da opinião que a soberania do Estado, em sentido vertical, tinha um limite, acima do qual o espaço era completamente livre." Luis Ivani de Amorim Araújo. *Curso de Direito Internacional Público*. 6. ed. Rio de Janeiro, Brazil: Forense, 1988. p. 198.

Marco G. Marcoff defended the provision of a conventional vertical limit, based on scientific criteria, but supported by the concern of distinguishing the legal regimes of Air Law and Space, in an altitude around 80–200 km from sea level.[104]

The 80 km standard was also suggested by Maurice N. Andem, who endorsed the idea that this regulation should be provided by a treaty dealing also with the passage of spaces objects below or above such altitude. Andem concluded that the arbitrary delimitation could guarantee, to all States, equal conditions to defend their air space, while contributing to the development of civil aviation and strengthening the principles of outer space's freedom of exploration and peaceful use.[105]

Proposals regarding a low air space vertical limit must be acknowledged, since they take into consideration the maximum altitude reasonably achievable by commercial aircrafts. At the beginning of space age, Albert Moon proposed that State sovereignty should rise until 5 miles (8,045 km) of altitude, in an adaptation to Space Law of the Maritime Law thesis that the territorial sea ended at 3 nautical miles (5,484 km) off the coast, as advocated by several nations.[106] In a recent work presented to the McGill University, Dean Reinhardt revisited the possibility of a low limit, suggesting the altitude of 12 nautical miles (22,224 km) from sea level, therefore applying the same standard provided by the Montego Bay Convention of 1982 regarding the territorial sea. For the author, the provision of a low altitude vertical limit would assure that outer space could be accessed by every State.[107]

Among the arbitrary delimitation critics, Myres S. McDougal, Harold D. Lasswell and Ivan I. Vlasic stand out as affirming that those who defended the creation of a vertical limit, at low or high altitude, will always fail in providing a rational accommodation of the inclusive and exclusive interests of States.[108] This is a challenge that must be recognized and duly taken into consideration.

---

[104] Marco G. Marcoff. *Traité de Droit International Public de l'Espace*. Fribourg, Switzerland: Éditions Universitaires Fribourg, 1973. p. 325.

[105] Maurice N. Andem. *International Legal Problems in the Peaceful Exploration and Use of Outer Space*. Rovaniemi, Finland: University of Lapland, 1992. p. 152.

[106] Albert Moon. "A Look at Airspace Sovereignty". *Journal of Air Law and Commerce*, n. 29. Dallas, USA, 1963. p. 328.

[107] "Space tourism and other vehicles capable of operating in the near space area are nearing commercial feasibility. Intercontinental hypersonic vehicles are being planned. A low limit on State sovereignty would allow these vehicles to operate freely without being obstructed by a political veto from the underlying State. Overflight rights would not be required which would save all the time an effort required to negotiate the web of air transit agreements now required to operate an international airline. More and more States are developing their own domestic space launch capability. Few of these new space powers will be able to freely access space, or utilize the most efficient launch azimuths, if neighboring States can claim sovereignty up to even 62 miles (100 km). They will have even more difficulty returning objects to Earth if the boundary is set at that altitude. Even the U.S. and Russia are facing limitations on their ability to freely access space. Setting a low vertical limit on State sovereignty will ensure all States have equal access to space." Dean Reinhardt. *The Vertical Limit of State Sovereignty*. Institute of Air and Space Law, McGill University. Montreal, Canada: 2005. p. 75.

[108] "Any such arbitrary chosen boundaries must again obviously fail to provide rational accommodation between inclusive and exclusive interests". Myres S. McDougal, Harold D. Lasswell and Ivan I. Vlasic. *Law and Public Order in Space*. New Haven, USA: Yale University Press, 1963. p. 349.

Chapter 5
The "Protozone"/"Mesospace" Situation

> Since the upper limit of aerodynamic lift may not coincide with the lower limit of outer space, some say there is an intermediate zone between the two, which most of those then call mesospace. (Peter P. C. Haanappel), The Law and Policy of Air Space and Outer Space[1]

Above the altitude routinely used by commercial aviation and below the lowest orbit manageable by a space object, there is a gray area where aeronautic and astronautic activities may intersect each other. With the constant evolution of technology, this region has increasingly received more attention by legal scholars, international organizations and delegations before the UNCOPUOS, in efforts to properly regulate it through a common international regime.

Joseph N. Pelton assigned the name "protozone" to this area, above 21 km and below 100 km of altitude, identifying it as an unregulated zone above commercial air flight and below traditional definitions of outer space. In such region, near space technologies and suborbital flights present relevant issues for the international community, thus demanding an applicable regulatory framework.[2]

An interesting perspective of the spatialist approach suggests the segregation of air space and outer space in at least three distinctive zones, creating a so-called "mesospace", to be subjected to legal regime similar to the applicable to navigation through the territorial sea and the contiguous zone, where passage rights would be effective. As indicated by Adherbal Meira Mattos, consequently "there would be the air space, where the subjacent State would hold complete sovereignty, the

---

[1] Peter P. C. Haanappel. *The Law and Policy of Air Space and Outer Space: a Comparative Approach*. The Hague, the Netherlands: Kluwer Law International, 2003. p. 27.

[2] Joseph N. Pelton. "A New Integrated Global Regulatory Regime for Air and Space: The Needs for Safety Standards for the 'Protozone'". *2nd Manfred Lachs International Conference on Global Space Governance*. Montreal, Canada: McGill University, 2014.

© Springer International Publishing Switzerland 2015

O.O. Bittencourt Neto, *Defining the Limits of Outer Space for Regulatory Purposes*, SpringerBriefs in Space Development,
DOI 10.1007/978-3-319-16685-8_5

contiguous space, where it would exercise relative sovereignty and the outer space, where there is no exercise of sovereignty."[3]

This alternative is often centered on the same basic considerations: (1) the limit to exclusive sovereignty of States should be located in a reasonably low altitude, in order to allow the free exploration and use of outer space; (2) an intermediate zone, contiguous to the first one, should be established in such a way as allowing the territorial State to preserve residual and limited jurisdiction for based on national security concerns, but without the right to prohibit the innocent passage of foreign space objects; finally, (3) outer space, free from national claims, should only start where the intermediate zone ended.[4]

It is possible to identify A. Merignac, a follower of the teachings of Paul Fauchille, as the pioneer on the delimitation of vertical space in separate zones. In accordance with Marignac's ideas, presented just before the beginning of the First World War, States should hold sovereignty up to 200 m of altitude, above which passage rights should be applicable until a level of 400 m, where freedom would prevail.[5]

As remarked by Luis Ivani de Amorim Araújo, such a rationale was espoused in Space Law by John C. Cooper, before the International Law Association (ILA), in its session held in Washington, 1956. At that occasion, Cooper proposed the division of vertical space in three zones: the first one, under absolute sovereignty of States, should be limited to the maximum height an aircraft could reach; the second, named "contiguous space", would go up to 300 miles (482.7 km) of altitude, permitting freedom of passage for non-military objects; finally, above it, outer space should be found, free from national claims.[6]

Without losing his belief on the division of zones above the ground, John C. Cooper later provided a revision of his measurements, after realizing that satellites are capable of orbiting at altitudes well below the original 300 miles benchmark. Therefore, he proposed a new limit to "territorial space", extending itself until 25 miles (40.225 km) of altitude, above which should lie the "contiguous space" up to 75 miles (120.675 km) or, alternatively, up to the scientifically proved lowest altitude for orbital trajectories.[7]

Others formulated different versions of such reasoning. To Olgierd Wolczek, an intermediary space should be created that is located from 200 km of altitude to a superior standard to be later determined based on scientific data, where outer space

---

[3] "Teríamos um espaço aéreo, onde o Estado subjacente exerce soberania plena, um espaço contíguo, onde ele exerce uma soberania relativa e o espaço exterior, onde não há exercício de soberania." Adherbal Meira Mattos. *Direito Internacional Público*. 2. ed. Rio de Janeiro, Brazil: Renovar, 2002. p. 219.

[4] Nicolas Mateesco Matte. *Aerospace Law*. Toronto, Canada: Carswell, 1969. p. 39.

[5] A. Mérignac. "Le Domain Aérien Privé et Public et le Droit de l'Aviation en Temps de Paix et Guerre". *Revue Générale de Droit International Public*, Vol. XXI. Paris, France, 1914. p. 205 e ss.

[6] Luis Ivani de Amorim Araújo. *Curso de Direito Internacional Público*. 6. ed. Rio de Janeiro, Brazil: Forense, 1988. p. 211.

[7] John C. Cooper. *Explorations in Aerospace Law*. Montreal, Canada: McGill University Press, 1968. p. 304.

would commence.[8] Modesto Seara Vazques supported the delimitation in three zones, the first being related to air space, whose vertical limit could eventually be provided by treaty; the second would be devoted to the contiguous space, and should start where the first one ended and rise until 36,000 km of altitude, that is, up to the geostationary orbit, constituting a zone where every non-military object would preserve passage rights, as long as always in movement; finally, the third one would contemplate complete freedom of navigation, including for intelligent living beings that accepted such legal regime, by formal agreement.[9]

Following the same trend, Martin B. Schofield proposed the creation of three zones: the "national air space", until 50 miles (80.45 km) of altitude; the "international space", between 50 and 2,000 miles (80.45 and 3,218 km) of altitude; and finally the "outer space", above 2,000 miles (3,218 km) of altitude.[10]

It should also be mentioned the proposal of William A. Hyman regarding the creation of "Neutralia", a neutral zone between air space and outer space, where the innocent passage of any object would be authorized, without the territorial State having right to attack it or destroy it during its trajectory.[11] To Hyman, the "Neutralia" would be equivalent in Space Law to the Maritime Law concept of territorial sea, so that the sovereignty of the territorial State would, in no way whatsoever, be offended by aircraft and space objects while in such zone, even military craft.[12] It should be located between 50 and 150 miles (80.45–241.35 km) above sea level, but subjected to further revision, due to the possible development of space activities.[13] Interestingly enough, Hyman did not provide relevant differences between "Neutralia" and outer space, which, according to his proposal, would follow a legal regime if not identical, at least very similar to the former.

At the UN, the French delegation in 1959 stated that it would be possible to provide an intermediary and international zone amidst air space and outer space, where the exercise of sovereignty by the subjacent State would be limited.[14] Eventually, in

---

[8] Olgierd Wolczek. "Remarks". *Proceedings of the First Colloquium on the Law of Outer Space.* IISL, The Hague, 1957.

[9] Modesto Seara Vázquez. "The Functional Regulation of the Extra-Atmospheric Space". *Proceedings of the Second Colloquium on the Law of Outer Space.* IISL, London, 1958. p. 144/145.

[10] Martin B. Schofield. *Control of Outer Space.* New York, USA: 1959. p. 365.

[11] William A. Hyman. *Magna Carta of Space.* Amherst, USA: Amherst Press, 1962. p. 199.

[12] The publicist provided the following definition of the "Neutralia": "that there be established a neutral zone embracing the upper limits of airspace and the lower limits of outer-space to be known as 'Neutralia' in which the right of innocent passage shall be recognized without offense to sovereignty. In this area of 'Neutralia' there shall be the right of innocent passage of all craft, vehicles and objects of transit and movement without any such incident being deemed an invasion of sovereignty. In the event of any such innocent passage no nation shall have the right to attack or destroy any occupants thereof without prior, sufficient warning and notice of claim of invasion of sovereignty and without prior opportunity for determination of the merits of such complaints by peaceful methods". William A. Hyman. *Magna Carta of Space.* Amherst, USA: Amherst Press, 1962. p. 202.

[13] William A. Hyman. *Magna Carta of Space.* Amherst, USA: Amherst Press, 1962. p. 203/205.

[14] A/AC.98/C.2/SR.3.

1967, Sweden presented proposal based on the recognition of vertical layers for air space and outer space, but without indicating altitudes.[15]

An intermediary zone dividing air space and outer space was also supported by Bernard Heraud, where freedom of transit and innocent over flight would prevail, but moderated by the prohibition of activities that might be considered disadvantageous to the territorial State, who would then hold the burden of proving damages. To Heraud, "such a legal system would inject progressive evolution of the air law system into the outer space system, thus 'defanging' the problem of delimitation of the domains. One could speak in terms of a slow evaporation of national sovereignty in space." However, the vertical limits of such intermediary zone were not suggested by the author, who reasoned in favor of providing those standards arbitrarily, only marginally considering scientific features.[16]

The term "mesospace" was adopted, arguably for the first time, by Gijsbertha Cornelia Maria Reijnen, who identified it with the space located between 50 to 100 km of altitude above the sea level, as a region open to free passage of space objects and rockets, ruled by Space Law, with further regulation regarding its use and exploration to be eventually provided by States, through treaty.[17] Later, Peter P. C. Haanappel adopted the term mesospace in his comparative studies about Air Law and Space Law, proposing it as an intermediary zone amidst 40–90 km of altitude where passage rights would be applicable; its borders, however, should be under constant revision in order to reflect the development of aeronautical and space technology.[18]

At any rate, the proposals for division of the vertical space in successive zones, above the territory of States, creating an intermediary layer subjected to partial sovereignty rights in between the ones applicable to air space and outer space, find some prodigious challenges. Indeed, whenever those suggestions distinguish the location of each zone based on criteria previously addressed (i.e., atmospheric limit, aerodynamic lift, lowest perigee, gravitational effect etc.), they sustain similar complications. Besides, every time arbitrary standards for delimitation are favored, they require at least twice the effort when compared to the advantages of a sole vertical limit between air space and outer space. In fact, those proposals tend to consent with passage rights, for launchings and return of space objects, only as far as the intermediary zone is concerned. This inevitably implies that, below such layer, any trajectory through air space would be considered an invasion and violation of the territorial State's national integrity.

---

[15] A/AC.105/C.1/SR.47.

[16] Bernard Heraud. "The Problem of the Delimitation of the Outer Space". *Proceedings of the Twenty-Second Colloquium on the Law of Outer Space*. IISL, Munich, 1979.

[17] Gijsbertha Cornelia Maria Reijnen. *Legal Aspects of Outer Space*. Utrecht, the Netherlands: Drukkerij Elinkwijk, 1977. p. 86/91.

[18] Peter P. C. Haanappel. "Airspace, Outer Space and Mesospace". *Proceedings of the Nineteenth Colloquium on the Law of Outer Space*. IISL, New York, 1977. p. 161/162.

Chapter 6
National Legislation and Comparative Law

> Comparative Law is rising up over its horizons: a new presence is dispelling the shadows of the past to reveal a submerged bulk of buried cultural secrets. (Pier Giuseppe Monateri), Methods of Comparative Law[1]

The vertical limit of national sovereignty, in relation to the air space or outer space, was addressed by some rare municipal rules that conflict with one another.[2] Thus, it is import to resort to Comparative Law to achieve a comprehensive verification of those few alternatives presented by States for solution of the problem, in order to identify not only legal patterns but also to understand the reasons behind those laws. As taught by Rudolf Schlesinger:

> Unlike most other subjects in the Law school curriculum, Comparative Law is not a body of rules and principles. It is primarily a method, a way of looking at legal problems, legal institutions, and entire legal systems. By the use of the method of comparison, it becomes possible to make observations and to gain insights that would be denied to one whose study is limited to the law of a single country.[3]

As explained by Jacob Dollinger, "Comparative Law is the Science (or the method) of studying, by comparison, two or more legal systems, addressing their positive rules, their sources, their history and various social and political facts that influenced them."[4] The author concluded that Comparative Law does not have a

---

[1] Pier Giuseppe Monateri (ed.). *Methods of Comparative Law.* Cheltenham, UK: Edward Elgar Publishing, 2012. p. 1.

[2] C. Jayaraj. "Is There a Need for a Comprehensive Convention on Outer Space Law?" *Proceedings of the Fiftieth Colloquium on the Law of Outer Space.* IISL, Hyderabad, 2007.

[3] Rudolf B. Schlesinger, Hans W. Baade, Peter E. Herzog and Edward W. Wise. *Comparative Law.* 6. ed. New York, USA: 1998. p. 2.

[4] "O Direito Comparado é a ciência (ou o método) que estuda, por meio de contraste, dois ou mais sistemas jurídicos, analisando suas normas positivas, suas fontes, sua história e os variados fatores sociais e políticos que os influenciam." Jacob Dolinger. *Direito Internacional Privado – Parte Geral.* 5. ed. Rio de Janeiro, Brazil: Renovar, 2000. p. 40.

© Springer International Publishing Switzerland 2015

O.O. Bittencourt Neto, *Defining the Limits of Outer Space for Regulatory Purposes*, SpringerBriefs in Space Development,
DOI 10.1007/978-3-319-16685-8_6

positive set of rules, nor does it create legal rules. Rather it constitutes a scientific or methodological realm.[5]

Resorting to Comparative Law permits a study of the proposals accepted nationally by States in order to identify possible standards, as well as to verify their effects on their own legal system. By isolating similarities and differences, it may be possible to reach a deeper understanding of the subject matter and justify a proposal to, preferably, solve the current impasse.

It shall be recalled, as reminded by Francisco Rezek, that unilateral acts of States, including their national legislation, are recognized as sources of International Law, even though not included in the list encompassed by article 38 of the Statute of the International Court of Justice. As affirmed by the Brazilian law scholar, former judge of the International Court of Justice, an unilateral normative act "may eventually face outside, in its object, claiming the quality of source of International Law when capable of being referred by other States in support of a certain claim, or as justification of the legality of a certain action. Such is the case of laws and decrees by which each a State identifies, observing applicable limits, the extension of their territorial sea or exclusive economic zone, seaports regulations or even the authorization of interior waters' use by foreign vessels".[6] Indeed, Malcolm N. Shaw also stated that, "in certain circumstances, the unilateral acts of States, including statements made by relevant State officials, may give rise to international obligations."[7]

As pointed out by Dean Reinhard, even though some states have introduced legislation about Air Law after the Chicago Convention of 1944, dealing with the delimitation of the air space under their jurisdiction, there is no international consensus regarding the vertical limit to national sovereignty.[8]

An important contribution to the topic was provided by Australia, which unilaterally established that its national space activities occur or may occur above 100 km of altitude, consequently implying a vertical limit of national sovereignty.[9] However, the Australian delegation later reported to the UNCOPUOS that "the Act still does

---

[5] Jacob Dolinger. *Direito Internacional Privado – Parte Geral*. 5. ed. Rio de Janeiro, Brazil: Renovar, 2000. p. 43.

[6] "O ato normativo unilateral pode casualmente voltar-se para o exterior, em seu objeto, habilitando-se à qualidade de fonte de Direito Internacional na medida em que possa ser invocado por outros Estados em abono de uma vindicação qualquer, ou como esteio da licitude de certo procedimento. Tal é o caso das leis e decretos com que cada Estado determina, observados os limites próprios, a extensão do mar territorial ou da zona econômica exclusiva, o regime de seus portos, ou ainda a franquia de suas águas interiores à navegação estrangeira." Francisco Rezek. *Direito Internacional Público*. 11. ed. São Paulo, Brazil: Saraiva, 2008. p. 136.

[7] Malcolm N. Shaw. *International Law*. 5. ed. Cambridge, UK: Cambridge University Press, 2003. p. 114. Reference is made to relevant international case law in that regard: Legal Status of Eastern Greenland, Norway vs. Denmark, Permanent Court of Justice, 1933, Ser. A/B, No. 53, 71; Nuclear Tests Case, Australia and New Zealand vs. France, International Court of Justice, 1974, 253, 457.

[8] "In the years since the drafting of the Chicago Convention, States have taken different positions on the extent of vertical sovereignty and definitions of their national airspace. There is no consensus today." Dean Reinhardt. *The Vertical Limit of State Sovereignty*. Institute of Air and Space Law, McGill University. Montreal, Canada: 2005. p. 24.

[9] Space Activities Act, n. 34 1998-99, 1998.

not define "outer space" and the term is not defined in other Australian legislation. The 100-km altitude represents a practical clarification of where the Act applies. The 100-km altitude was not an attempt on Australia's part to define or delimit "outer space".[10]

It cannot be denied that the Australian Space Activities Act innovated in relation to such countries as Germany, which, when ruling about this subject, did not attempt to provide any clear indication of the boundaries between air space and outer space.[11]

Other States chose only to reaffirm sovereignty above their air space without, nevertheless, delimiting its maximum altitude. A relevant example can be found in the laws of the former USSR, later generally adopted by its successor, the Russian Federation.[12]

By the same token can be mentioned the Federal Constitution of Mexico, which was amended in 1960 to provide Mexican sovereignty in relation to the space above its territory up to the extension determined by International Law (article 42, VI).[13] Consequently, that legislation made reference to an international rule still to be provided by international Space Law.

The aerodynamic lift approach[14] found some acceptance among States, as far as their national legislation is concerned. For instance, the recent Austrian's Rule of Air, of 2010, in article 48, paragraph 2, "defines the upper state boundary as the height at which aircrafts can no longer operate by aerodynamic lift but only according to Kepler's laws".[15]

Recently, South Africa approved legislation regarding aviation and use of outer space that considered, at least apparently, the low standard of 18.5 km as the vertical limits of its national sovereignty.[16] Such a choice seems similar to Paul Fauchille's proposals of the early twentieth century.[17] Interestingly enough, a similar orientation

---

[10] A/AC.105/865/Add. 1.

[11] Air Navigation Law, 1964, art. 1(2).

[12] Aeronautic Code, 1961, art. 1.

[13] "El espacio situado sobre el territorio nacional, con la extensión y modalidades que establezca el propio derecho internacional."

[14] In accordance with the aerodynamic lift approach, the upper limits of our planet's atmosphere should, in one way or another, represent Air Law's jurisdiction, which is based on the premise that States have complete and exclusive sovereignty over the column of air above their territories. Beyond such a standard, outer space should commence, free from national claims.

[15] "Section 2 para. 48 of the Rule of the Air (Federal Law Gazette II No. 80/2010), a regulation implementing the Austrian Aviation Law, defines the upper state boundary as the height at which aircrafts can no longer operate by aerodynamic lift but only according to Kepler's laws." Official answer from Austria, presented in 2011, to a questionnaire prepared by UNCOPUOS Legal Subcommittee regarding national legislation about the definition and delimitation of outer space. A/AC.105/C.2/2011/CRP.10.

[16] Strategic Geographic Advantage Act, 2008.

[17] Paul Fauchille. *Le Domaine Aéreian et Le Régime Juridique de les Aerostats*. Paris, France: Pendone, 1901.

can be found in Belarus, which considers that its national "classified air space" is located up to 20.1 km of altitude, for national security concerns.[18]

One should also make reference to Regulation 428/2009 of May 5th, 2009, by the European Union Council, regarding export controls, that defined "space qualified products" as "products designed, manufactured and tested to meet the special electrical, mechanical or environmental requirements for use in the launch and deployment of satellites or high altitude flight systems operating at altitudes of 100 km or higher".

Occasional references to the vertical frontier may be found in municipal legislation addressing specific space activities. For instance, the USA, that has offered a coherent position before the UNCOPUOS according to which no delimitation or definition is needed at this moment, approved regulation by its Air Force establishing that "a USAF rated officer qualified to perform duties in space (50 miles above the earth's surface) who completes a minimum of one operational mission is eligible for the astronaut qualifier (pilot astronaut, CSO astronaut, observer astronaut, ABM astronaut, and flight surgeon astronaut)".[19]

It must be acknowledged that the term "space" is currently defined by the United States Code, in the chapter regarding fiscal matters, as "any area not within the jurisdiction (as recognized by the United States) of a foreign country, possession of the United States, or the United States, and not in international water".[20]

Additionally, a couple of American federal states have proposed or approved legislation defining the air space under their jurisdiction based on vertical limits, for instance, Virginia and New Mexico.[21] Such rules may eventually be challenged at the US Supreme Court, arguably due to violation of federal competence.[22]

---

[18] "Law No. 156-3 of 5 May 1998, on objects belonging exclusively to the State, declares that the airspace above the territory of Belarus is the exclusive property of the State. As regards the issue of the definition and delimitation of outer space, however, Belarus, which embarked on outer space activities only recently, does not yet have separate domestic legislation relating to outer space but is currently developing legislation that will, inter alia, cover that issue. Current law divides the airspace of Belarus into two categories: classified and unclassified. Airspace below an altitude of 20,100 m is classified and flights within it are governed by domestic legislation: the Air Code and the Rules for the Use of Airspace adopted by Order No. 1471 of the Council of Ministers on 4 November 2006. Outside classified airspace (above an altitude of 20,100 m), which is considered outer space, the provisions of international agreements apply." A/AC.105/865/Add.4.

[19] Instruction 11-42, 2003, item 2.2.

[20] http://uscode.house.gov/, accessed on 09.02.2012.

[21] Regarding the American State of Virginia, it was proposed that the term "suborbital flight" should be understood as those that take place up to 62.5 miles of altitude above mean sea level, during the preparation of amendments to the "Space Flight Liability and Immunity Act, of 2007, but the final wording did not incorporate such suggestion. http://leg1.state.va.us/cgi-bin/legp504.exe?071+sum+HB3184, accessed on 08.27.2012". On the other hand, the American State of New Mexico, in its "Gross Receipts and Compensating Tax", of 2007, defined outer space as anywhere above 60,000 feet of altitude from mean sea level. http://cfr.vlex.com/vid/52-229-gross-receipts-compensating-19871177, accessed on 08.27.2012.

[22] Francys Lyall and Paul B. Laursen. *Space Law: a Treatise*. Farnham, England: Ashgate, 2009. p. 160.

The new Federal Constitution of Ecuador, approved in 2008, claimed national sovereignty over the geostationary orbit's segment above its territory, suggesting a territorial limit of, at least, 36,000 km of altitude. In fact, its article 4 solemnly stated: "the Ecuadorian State shall exercise rights in relation to the corresponding segments of the geostationary synchronous orbit, maritime spaces and Antarctica."[23] In similar terms, the Colombian Federal Constitution of 1991, in its article 101,[24] contemplated identical declaration by including the geostationary orbit as part of the national territory, but with this interesting remark: as long as there are no international rules preventing it.[25] It is important to recall that both nations were part of the polemic Bogotá Declaration of 1976, by which equatorial States asserted that, since the Space Treaty of 1967 did not contemplate the delimitation of outer space, national appropriation of the geostationary orbit by the States underneath it did not violate International Law.[26] It is also important to stress that Brazil, before the UNCOPUOS, has not presented any official national claim in relation to the segment of geostationary orbit above its territory, as in fact performed by the likes of Ecuador and Colombia.[27]

Nandasiri Jasentuliyana opposed the equatorial States claims based on the article 2 of the Space Treaty of 1967, which forbids national appropriation of any part of outer space. Regarding rules applicable to allocation of telecommunication satellites and respective radio frequencies, Jasentuliyana reaffirmed that the International Telecommunications Unity (ITU) has jurisdiction over it and works in order to answer the general international interest.[28]

---

[23] "El Estado ecuatoriano ejercerá derechos sobre los segmentos correspondientes de la órbita sincrónica geoestacionaria, los espacios marítimos y la Antártida."

[24] "También son parte de Colombia, el subsuelo, el mar territorial, la zona contigua, la plataforma continental, la zona económica exclusiva, el espacio aéreo, el segmento de la órbita geoestacionaria, el espectro electromagnético y el espacio donde actúa, de conformidad con el Derecho Internacional o con las leyes colombianas a falta de normas internacionales."

[25] Jairo A. Becerra Ortiz. "A Survey of Colombia's New Outer Space Policy: Reforms in Colombian Law", *Acta Astronautica,* Vol. 63, n. 1-4. Washington, USA, July-August 2008. p. 560/563.

[26] By the "Bogotá Declaration", of December 3rd 1976, several equatorial States claimed sovereignty over the geostationary orbit, due to its strategic importance, particularly as far as telecommunications are concerned. Said instrument, in which Brazil took part of the negotiations but later denied being bound to, represented the position of several developing countries crossed by the Equator line. It was also signed by Colombia, Congo, Ecuador, Indonesia, Kenya, Uganda and Zaire.

[27] Consideration of the legal regime applicable to the geostationary orbit was included on UNCOPUOS permanent agenda in 1977, and, from the next year on, became a topic to be addressed the study group on delimitation and definition of outer space. For more information: A/32/20.

[28] Nandasiri Jasentuliyana. *International Space Law and the United Nations.* The Hague, the Netherlands: Kluwer, 1999. p. 53. Important to remark that the ITU, in its Regulation of Radio Waves, article S1.64, defines "space station" as "a station located on an object which is beyond, is intended to go beyond, or has been beyond, the major portion of the Earth's atmosphere". See also: Frans G. Von der Dunk. "The Sky is the Limit – but Where does it End?" *Proceedings of the Forty-Eighth Colloquium on the Law of Outer Space.* IISL, Fukuoka, 2005. p. 88.

The present scenario, based on Comparative Law methods, observing municipal legislations in search of a delimitation solution, shows that national legislators adopted different criteria to determine the vertical extension of their State's sovereignty. Those few that included, direct or indirectly, a clear boundary to national air space, vary from the ones that were only concerned with the reasonably low altitude where commercial aviation is conducted to those that extended national sovereignty up to the most valuable orbits, including the geostationary one.

In conclusion, one may infer that the municipal provision of the vertical limit of State sovereignty, whenever provided, leans towards local interests—which should come as no surprise, considering the strategic importance of aeronautic and space activities. Unless an international provision is agreed upon in that regard, it seems reasonable to foresee a future, complex scenario where the border of air space and outer space can differ from one country to another, rendering amazing challenges to international cooperation and hindering the progress of space activities. If the Law of the Seas may serve as an indicator, as far as national limits of the territorial sea are concerned, universal delimitation should always prevail upon local and unilateral solutions, for the benefit of peaceful international relations.

Chapter 7
Towards a Compromise

> Having established that outer space is a distinct dimension, possessing a special legal status, we must ask the question: where are its frontiers? (Manfred Lachs), The Law of Outer Space[1]

Spatialists and functionalists have been dueling for decades, and it is impossible to find a clear winner.[2] In a study regarding the position advocated by legal scholars on this topic, C. S. Tang conducted extensive research of the various proposals for delimitation of outer space. The author concluded that it was not possible to identify one thesis that received support overwhelming enough to be considered the preponderant one.[3] By its turn, a report prepared by the UN General Assembly in 2002, presenting a detailed summary of the debates, concluded that "no agreement on substantive legal issues relating to the definition and delimitation of outer space are apparent from the reports of the Subcommittee or of the Committee on the Peaceful Uses of Outer Space."[4] The stalemate is such that since 1987 some delegations to the UNCOPUOS have defended the withdrawal of this topic of the legal subcommittee's permanent agenda.[5]

During the 2007 session of the legal subcommittee, Brazil achieved the presidency of the study group devoted to the delimitation and definition of outer space, through delegate José Monserrat Filho. Among the first decisions adopted therein was the consultation of Member States regarding the necessity of studying the topic,

---

[1] Manfred Lachs. *The Law of Outer Space*. Leiden, the Netherlands: Martinus Nijhoff, 2010. p. 53.

[2] W. W. C. de Vries. "Boundaries in Space? Some Theories and Problems Regarding the Definition, Delimitation and/or Demarcation of Outer Space". *Thesauros Acroasium*. Institution of International Public Law and International Relations of Thessaloniki. Vol. XIV, 1985. Tessaloniki, Greece. p. 782.

[3] C. S. Tang. "The Boundary Question in Space Law: a Balance Sheet". *Ottawa Law Review*, Vol. 6, n. 263. Ottawa, Canada, 1973. p. 273.

[4] A/AC.105/769. p. 7.

[5] A/AC.105/769. p. 5.

© Springer International Publishing Switzerland 2015

O.O. Bittencourt Neto, *Defining the Limits of Outer Space for Regulatory Purposes*, SpringerBriefs in Space Development,
DOI 10.1007/978-3-319-16685-8_7

considering the state of the art of aerospace technology, as well as requesting proposals for solution of the current standoff between spatialists and functionalists.[6] Thus, it seems important to stress that this study does not constitute a mere academic exercise, since it seeks to contribute to a future positioning of the Brazilian delegation before the UNCOPUOS, as well as providing food for thought for national legislators.

Sovereignty has represented a key feature of International Law right from the beginning, and the respect afforded to the territorial integrity of fellow States represent a fundamental tenet of the field. Sovereign rights are applicable domestically, as superior power over the population of a certain territory, as well as internationally, regarding independence and equality among nations.[7] James Crawford indicated that "the term 'sovereignty' has a long and troubled history, and a variety of meanings. In its most common modern usage, sovereignty is the term for the 'totality of international rights and duties recognized by international law' as residing in an independent territorial unit – the State. It is not itself a right, nor is it a criterion of statehood (sovereignty is an attribute of States, not a precondition). It is a somewhat unhelpful, but firmly established, description of statehood; a brief term for the State's attribute of more-or-less plenary competence".[8]

Complete State jurisdiction, understood as the capacity to exercise every public function, encompassing jurisdictional, executive and adjudicatory counterparts, shall be circumscribed to the respective national territory, and must be provided in accordance with International Law.[9] As explained by Malcolm N. Shaw, "since such fundamental legal concepts as sovereignty and jurisdiction can only be comprehended in relation to territory, it follows that the legal nature of territory becomes a vital part in any study of international law. Indeed, the principle whereby a state is deemed to exercise exclusive power over its territory can be regarded as a fundamental axiom of classical international law."[10]

Therefore, areas of contact between domestic legal systems, i.e., the political frontiers, corresponding to the uttermost corners of national territories, should preferably be demarcated in order to avoid conflicts of jurisdiction.[11] In accordance with Victor Prescott and Gillian T. Triggs, "political frontiers once separated neighboring countries and geographical interests in them is mainly concerned with their physical characteristics, their position, the attitudes and policies of the flanking states, the

---

[6] A/AC.105/891.

[7] F. H. Hinsley. *El Concepto de Soberania*. Barcelona, Spain: Labor, 1975. p. 153/166.

[8] James Crawford. *The Creation of States in International Law*. 2. ed. New York, USA: Oxford University Press, 2006. p. 32.

[9] "Como a sociedade internacional é uma sociedade de justaposição, os Estados não dependem de qualquer outro poder político que lhes seja superior. Esta constatação deve ser colocada no âmbito da evolução desta sociedade. Embora soberanos, os Estados encontram-se submetidos ao direito internacional que, em princípio funda as suas competências respectivas." Paul Reuter. *Instituições Internacionais*. Lisbon, Portugal: Edições Rolim, 1975. p. 148.

[10] Malcolm N. Shaw. *International Law*. 5. ed. Cambridge, England: 2003. p. 409.

[11] Adherbal Meira Mattos. *Direito Internacional Público*. 2. ed. Rio de Janeiro, Brazil: Renovar, 2002. p. 65.

influence of the frontier on subsequent development of the cultural landscape and the way which boundaries were drawn within the frontier."[12]

It must be acknowledged that the existence of clearly delimited frontiers is not a prerequisite for international recognition of statehood.[13] However, at least since the USSR dissolution in 1991, the study of legal aspects related to frontiers and boundaries regained relevance in International Law, in order to justify control and clear identification of national borders, due to historical, political and economic concerns.[14]

Positive, universally agreed delimitation of territories should be encouraged in order to assure international stability and peaceful relations among States. As indicated by Ian Brownlie, "the legal competences of states and the rules for their protection depend on and assume the existence of a stable, physically delimited, homeland."[15] Without clear boundaries, especially as far as strategic locations are concerned, the potential of dangerous international disputes raises dramatically.[16]

Indeed, John Kish concluded that "considering the conflicting strategic regimes of territorial sovereignty and international spaces, international conflicts may arise from the military use of the disputed marginal zones of national spaces and international spaces. Moreover, considering the conflicting strategic regimes of various international spaces, international conflicts may arise from the military use of the disputed marginal zones of various international places. The potential threat of international conflicts arising from the absence of uniform international delimitation justifies the necessity of a generally accepted international regulation concerning the delimitation of national spaces and international spaces."[17] Therefore, a vertical limit to national sovereignty ought not only be supported, but bindingly provided, to assure peace and international security.

Delimitation by Treaty

The differences between the legal regimes applicable to air space and outer space are of a fundamental order: while Air Law is based on sovereign considerations, Space Law overtly forbids any form of national appropriation. To Harnam Bayhana, Air Law and Space Law cannot be correctly applied and interpreted without the

[12] Victor Prescott and Gillian T. Triggs. *International Frontiers and Boundaries: Law, Politics and Geography*. Dordrecht, the Netherlands: Martinus Nijhoff, 2008. p. 30.

[13] Antonio Cassese. *International Law*. 2. ed. Oxford, England: Oxford University Press, 2005. p. 73.

[14] Michel Foucher. *Obsessão por Fronteiras*. São Paulo, Brazil: Radical Livros, 2009. p. 21/27.

[15] Ian Brownlie. *Principles of Public International Law*. 6. ed. Oxford, England: Oxford University Press, 2003. p. 105.

[16] Michel Fouchet. *L'Obsession des Frontières*. Paris, France: Librairie Académique Perrin, 2007.

[17] John Kisch. *The Law of International Spaces*. Leiden, the Netherlands: Sijthoff, 1973. p. 51.

provision of a clear delimitation of their respective realms.[18] Consequently, the respective jurisdictions of such incommunicable systems must, once and for all, be clearly established.

The growing exploitation of outer space, especially by private parties, demands rational use of this environment, in order to respect the common interests of all nations. Manfred Lachs, former judge of the International Court of Justice (ICJ) who presided upon the works of UNCOPUOS for years, including the period when the most important Space Law treaties were conceived, asserted:

> (…), with the growth of activities in outer space such a delimitation would offer clear advantages. It would prevent the misunderstanding or even friction to which uncertainty tends to give rise, facilitate international cooperation.[19]

The delimitation of outer space is also justified for safeguarding one of the most important Space Law principles, that is, the prohibition of national appropriation over outer space, including orbital positions and celestial bodies. In a seminal study, Armand D. Roth indicated that such a rule could only be truly effective if and when the outer space environment becomes conveniently identified[20]; otherwise, claims could eventually be presented by States in relation to relevant outer space orbits and orbital positions, as demonstrated by the already mentioned Bogotá Declaration of 1976. Damodar Wadegaonkar noted that "instances like the Bogotá Declaration reveal a future trend, namely, that when States' national interests are jeopardize they would challenge the accepted norms of international Space Law, relying on the fact that no accepted or adequate definition of Outer Space exists. The international community has to muster up enough consensus to formulate principles on this subject so as to prevent such misconception, misapprehension as well as misguided claims in the future."[21]

Currently, due to the lack of a multilateral regime on that regard, it may be possible to envision a sad repetition of what happened in relation to national delimitation of the territorial sea, when, throughout the years, several States claimed different boundaries based on municipal legislation, randomly adopting criteria that only answered to their own local interests.[22] As explained by Antonio Cassese, "within the territorial sea (plus its airspace, seabed and subsoil) the coastal State enjoys full sovereignty, subject to the right of innocent passage of foreign merchants' ships and warships (however, for submarines it is required that they must navigate on the

---

[18] Harnam Bhayana. *International Law in the Regime of Outer Space*. Calcutta, India: R. Cambray & Co. Pvt., 2001. p. 142.

[19] Manfred Lachs. *The Law of Outer Space*. Leiden, the Netherlands: Martinus Nijhoff, 2010. p. 5.

[20] "Le danger de revendications sur l'espace extra-atmosphérique ne saurait être exclu en l'absence de délimitation de ce dernier. Le principe de non-appropriation ne déploie ses effets que dans un cadre spatial qu'il convient de préciser". Armand D. Roth. *La Prohibition de l'Apropriation et les Regimes d'Accès aux Espaces Extra-Terrestres*. Paris, France: Presses Universitaires de France, 1992. p. 93.

[21] Damodar Wadegaonkar. *The Orbit of Space Law*. London, UK: Stevens & Sons, 1984. p. 45.

[22] Paul Reuter. *Droit International Public*. 4. ed. Paris, France: Presses Universitaires de France. 1973. p. 247/251.

surface and show their flag)."[23] It was only with the Montego Bay Convention of 1982 that the international system of Law of the Seas was finally ordered, despite the position of a fading group of States that still claim a territorial sea beyond the international prescription of 12 nautical miles.[24]

The irresistible evolution of technology has presented new challenges to Space Law that could not be anticipated when the space treaties were drafted. Concluded Walter W. C. de Vries: "we shall bear in mind that the 1967 Outer Space Treaty, although raised to a level of a 'Magna Carta', may not suffice for the future. As our normative thinking is continually changing and gradually emerging within the legal process, new principles may arise which might even be contrary to these enumerated in the 1967 Outer Space Treaty."[25] Important questions on the definition and delimitation of outer space still await a proper solution, demanding the attention of the international community.

As far as international liability for space objects is concerned, the 1972 Convention contemplated a double standard that depends upon the place where the damage occurs. If caused on the surface of the Earth or to an aircraft in flight, the Launching State shall bear absolute liability, hence answering in all situations, even for force majeure.[26] On the other hand, if the damage is produced by a space object elsewhere than on the surface of the Earth to other space objects or to persons or property on board such space objects, liability shall depend on the occurrence of fault.[27] As it can be noticed, there is an intrinsic link with the "locus in quo", i.e., the place of damage, for determination of the liability regime concerning a certain damage produced by space objects, what, by itself, justifies efforts regarding the delimitation of air space/outer space "border".[28] The Brazilian delegation to the UNCOPUOS considered such scenario in 2009, supporting the arguments presented by José Monserrat Filho as president of the working group concerned with delimitation and definition of outer space.[29]

---

[23] Antonio Cassese. *International Law*. 2. ed. Oxford, England: Oxford University Press, 2005. p. 85.

[24] Nguyen Quoc Dinh, Patrick Daillier and Alain Pellet. *Direito Internacional Público*. Lisbon, Portugal: Fundação Calouste Gulbenkian, 1999. p. 1183.

[25] Walter W. C. de Vries. "The Creation of a Concept of the Law of Outer Space", *In:* Tanja Masson-Zwaan (ed.). *Space Law: Views of the Future.* Deventer, the Netherlands: Kluwer, 1988. p. 31.

[26] "A launching State shall be absolutely liable to pay compensation for damage caused by its space object on the surface of the Earth or to aircraft in flight".

[27] "In the event of damage being caused elsewhere than on the surface of the Earth to a space object of one launching State or to persons or property on board such a space object by a space object of another launching State, the latter shall be liable only if the damage is due to its fault or the fault of persons for whom it is responsible".

[28] "Elle [la Convention de 1972] institue deux régimes différents de responsabilité ayant trait aux activités qui se déroulent dans l'espace extra-atmosphérique, en fonction de la localisation du dommage considéré. Elle distingue en cela les dommages occasionnés à autri à la surface e la Terre et ceux qui sont ocasionnés dans l'espace ou sur un corps céleste." Jean-Paul Pancracio. *Droit International des Espaces*. Paris, France: Armand Colin, 1997. p. 241.

[29] A/AC.105/935.

There remains no doubt that the unstoppable evolution of technology is progressively narrowing the differences between aeronautic and astronautic activities. I. H. Ph. Diedericks-Verschoor, while supporting the necessity of delimitation, stated:

> It would be quite wrong to think that demarcation in space is largely a matter for academic debate. On the contrary, a practical solution is required with ever increasing urgency.[30]

The development of hybrid vehicles like the American space shuttle present relevant questions to Space Law. Even though taking off as rockets, they return to the Earth as gliders. Their missions are concerned with uses of outer space, despite employing wings, appendices and other aerodynamic devices similar to those used by aircrafts. In accordance with the functionalist approach, there would not be grounds for States to claim violation of their air space during launching or return of space shuttles, irrespective of their location at any given time, since their sole purpose is the performance of space activities.[31]

That does not seem to be the best solution, since during overflight of foreign air space, aerospace vehicles present relevant dangers to the local air traffic and to regular commercial flights. Nothing justifies a privileged treatment to an object that, at least during part of its journey, falls into the definition of aircraft applicable to Air Law, since being a "machine that can derive support in the atmosphere from the reactions of the air other than the reactions of the air against the Earth's".[32]

Also, the current projects of "aerospace planes" must be taken into consideration. These vehicles will take off and land as an aircraft, relying on aerodynamic lift for part of their courses, but at same time being capable of reaching the upper atmosphere layers in order to arrive to their destiny in a much faster than standard commercial airplanes.[33] Ruwantissa I. R. Abeyratne envisioned a broad definition of

[30] "It would be quite wrong to think that demarcation in space is largely a matter for academic debate. On the contrary, a practical solution is required with ever increasing urgency". I. H. Ph. Diederiks-Verschoor. *An Introduction to Space Law*. 2. ed. Dordrecht, the Netherlands: Kluwer Academic Publishers, 1999. p. 17/18.

[31] "The Space Shuttle has both military and civilian functions. It can carry a total of seven people (crew and passengers) for a period of up to thirty days. One of its functions is to provide a platform for a space tug to put a military assistant satellite into deep orbit. It can also be used to refuel other spacecraft, to rotate crews on long range missions, or to scan the airspace for close-in observation. The shuttle's primary function however, is to provide low cost transportation to and from Earth orbit, and its first operational mission was to transport space-labs." I. H. Ph. Diederiks-Verschoor. *An Introduction to Space Law*. 2. ed. Dordrecht, the Netherlands: Kluwer Academic Publishers, 1999. p. 86.

[32] Chicago Convention, 1944, Annexes II, VII and XI.

[33] "The term "aerospace plane" reflects two important features of these objects. They will be constructed, using both aeronautical and space technologies, and they would be able of flying both in airspace and outer space. As aerospace technology continues to develop, air and space lawyers will find themselves confronted with new issues. Many well-established concepts will require reassessment, in light of new technical developments." Vladen Vereshchetin. "New Steps in International Space Law". *In* Nandasiri Jasentuliyana. *Perspectives in International Law.* The Hague, the Netherlands: Kluwer, 1995. p. 469.

those devices: "an aerospace plane is a hypersonic single stage to orbit reusable vehicle that takes off and lands horizontally on a conventional runway. (…) The aerospace plane will be constructed with the use of aeronautical and space technologies and would be capable and, indeed, required to fly both in air space and outer space. In these circumstances, the most fundamental issue at hand is the applicability of laws appropriate to the space plane's activity."[34]

According to the functionalist approach, those objects would be considered an aircraft when used for transport between land destinations, even though during part of their voyage they may reach altitudes as high as or even above 100 km from mean sea level.[35] Recently, the USA concluded unmanned tests of military space planes, including the X-37B[36] and the Falcon HTV-2,[37] which impose particular concerns since they would be considered a type of state aircraft by the Chicago Convention of 1944, therefore not being granted any sort of *a priori* passage rights through foreign air space.

The matter gets even more complicated in relation to sub-orbital vehicles currently under development for exploitation of the multimillionaire space tourism market.[38] In general, the more advanced projects encompass the taking off capability of an aircraft, and then a second stage where a small section is detached from the main vehicle and ejected to high altitudes by rockets, only to later return to the ground as a glider, controlling its fall using aerodynamic devices.

That was the system adopted, in 2004, by the first private spaceship designed for space tourism, named SpaceShipOne, which managed to achieve 100 km of altitude in different flights conducted during 2 weeks, awarding its project managers, led by the aerospace engineer Burt Rutan, the Ansari X-Prize.[39]

The International Civil Aviation Organization (ICAO) considered the suborbital flights topic in a report presented at its Council's 175th meeting, on May 30th, 2005, which defined it as "a flight up to a very high altitude which does not involve

---

[34] Ruwantissa I. R. Abeyratne. *Frontiers of Aerospace Law.* Burlington, USA: Ashgate, 2002. p. 22.

[35] S. Bhatt defended the creation of an aerospace legal regime to deal with future space plane flights. *International Aviation and Outer Space Law and Relations: Reflections on Future Trends.* New Deli, India: Asian Institute of Transport Development, 1996. p. 135/139.

[36] http://www.msnbc.msn.com/id/40491145/ns/technology_and_science-space/t/secretive-x-b-robot-space-plane-returns-earth/, accessed 09.26.2011.

[37] http://www.dailymail.co.uk/sciencetech/article-2032679/First-footage-Falcon-HTV-2-hypersonic-spacecraft-test-flight-released.html, accessed on 09.26.2011.

[38] Recently, the International Institute of Air and Space Law (IIASL) organized the "Symposium on the Regulation of Sub-Orbital Flights in the European Context", on 09.16.2010, in Leiden, the Netherlands. For more information about the presentations, see Leiden University website: http://law.leiden.edu/organisation/publiclaw/iiasl/conferences/sub-orbital-space-tourism-regulation-in-the-european-context.html, accessed on 02.03.2011.

[39] Regarding the success of the *SpaceShipOne* and its later acquisition by the Virgin Group, that intends to use an upgraded version, named *SpaceShipTwo,* for commercial suborbital flights in the near future, see: Kenny Kemp. *Destination Space: How Space Tourism is Making Science Fiction a Reality.* London, England: Virgin Books, 2007.

sending the vehicle into orbit."[40] The document acknowledged that SpaceShipOne was registered in the USA as a rocket, but not before the UN as a space object, apparently because such register would only be required for objects launched to Earth orbit or beyond, as provided by the article 2 of the Registration Convention of 1975, and "not to objects performing sub-orbital flights per se."[41]

The referred ICAO report concluded that:

> From a spatialist viewpoint, there is no clear indication in international law on the delimitation between airspace and outer *space* which would permit to conclude on the applicability of either air law or space law to sub-orbital flights. On the other hand, it might be argued from a functionalist viewpoint that air law would prevail since airspace would be the main centre of activities of sub-orbital vehicles in the course of an earth-to-earth transportation, any crossing of outer space being brief and only incidental to the flight.[42]

However, the functionalist approach could also allow another position, favorable to Space Law, due to the clear objective of the owners of said sub-orbital vehicle to provide, to their future clients, a "space tourism experience". After all, the marketing strategy of those companies highlight that their services provide access to outer space to anyone who can pay for their still quite pricey tickets.

Thus, since SpaceShipOne was registered in the USA as a rocket, it could be defended based on the functionalist rationale that an eventual violation of foreign air space should be ruled by Space Law and, therefore, no other State would have the right to exercise jurisdiction over those vessels, including through municipal legislation regarding national security or fiscal matters.

As indicated by Frans G. von der Dunk, American legislators chose to create an intermediary legal regime in the Commercial Space Launch Act of 2004, supporting private space endeavors like the SpaceShipOne without, nevertheless, modifying the traditional national philosophy which considers premature the delimitation of the air space/outer space boundary. Questioned von der Dunk: "is this current approach going to remain sufficient? The United States government may well have avoided the establishment of any boundary between air space and outer space, creating what amounts to a functionally-based regime for space tourism whilst using its territorial jurisdiction over the launch site, but might work satisfactorily only as long as such launches would take place exclusively from US territory. Launches conducted from elsewhere overflying US territory, once becoming seriously possible, could no longer be legally controlled in that way as long as they would not fly at such a low altitude that no one would contest this to be US air space, that is below a zone where it will be difficult or even impossible for a spacecraft to operate".[43]

Due to all of the above, it is believed that the spatialist approach is correct in its defense for provision of a clear delimitation between air space and outer space;

---

[40] C-WP/12436, 1.2.

[41] C-WP/12436, 3.3.

[42] C-WP/12436, 6.2.

[43] Frans G. Von der Dunk. "The Sky is the Limit – but Where does it End?" *Proceedings of the Forty-Eighth Colloquium on the Law of Outer Space*. IISL, Fukuoka, 2005. p. 91.

however, it must also be acknowledged that regulation regarding passage rights for launching and return of space objects ought also be addressed.

It is not ignored that several technical criteria were proposed for distinguishing the air space environment to the outer space one, without overwhelming success. In fact, the constant scientific evolution made many important law scholars change their minds, time and time again, in relation to the best approach to the spatialist rationale.[44] Therefore, the best alternative seems to be the adoption of an arbitrary delimitation of State sovereignty, preferably by treaty, that includes regulation regarding passage rights for space activities through foreign airspace, considering the relevant public concerns of the overflown countries.

The preference hereby manifested in favor of providing such rule via treaty is justified by the intent of assuring the utmost legal security possible, in order to avoid possible international disputes between Launching States and overflown States. If those rules were to be implemented by a non-binding international instrument, for instance, a UN General Assembly resolution, which constitutes what is commonly called "soft law", certainly an international custom could be developed based on the opinion and practice of States. Nevertheless, such customary rule could be quite difficult to prove, if ever formed. Besides, an eventual breach of a "soft law" does not necessarily imply disrespect to International Law, arguably justifying selective behavior based exclusively on national interests.[45]

Indeed, only through a multilateral treaty would it be possible to solve the problem herein presented in a definitive way, by a clear and effective regulation, discussed and approved by the international community in a common effort.

The Legal Standard of 100 km Above Sea Level

After recognizing that the lack of delimitation has not impeded the development of an orderly regime for the exploration and use of the space environment, Carl Q. Christol noted that "it may be timely to turn serious attention to the issue of delimitation/definition of outer space and associated boundary problems including the

[44] Bin Cheng, for instance, who originally defended that the air space should cover the limits of the atmosphere, later changed his mind, defending an arbitrary delimitation of the frontier between air space and outer space, through treaty. *Studies on International Space Law*. Oxford, England: Clarendon Pr, 1998. p. 448.

[45] Conforme indica Guido Fernando Silva Soares, "a oposição soft law versus hard law indicaria um contraste entre duas realidades coexistentes e que se auto-aplicam: tanto se encontra presente o fato tempo (a hard law seria um produto acabado, ao final de uma evolução geracional ao longo do tempo, portanto, a norma terminada em sua inteireza, e soft seria um vir a ser, um ato de potência, um ato de vontade dos Estados, que aspira a tornar-se uma norma), quanto o fator finalidade (na hard law, os Estados estabelecem obrigações jurídicas fortes, para serem efetivamente cumpridas, e na soft law existem normas jurídicas, mas seu cumprimento é meramente recomendado aos Estados, que podem, inclusive, não cumpri-las, sem que haja sanções aplicáveis aos inadimplentes." Guido Fernando Silva Soares. *Curso de Direito Internacional Público*. São Paulo, Brazil: Atlas, 2002.

geostationary orbit".[46] Decidedly, it shall be recognized that such delimitation can provide the legal security necessary to avoid international disputes regarding aeronautical and space activities, especially in relation to the use of spy satellites and to the safety of national air spaces eventually crossed by space objects during their launch or reentry.

That is the opinion of I. H. Ph. Diedericks-Verschoor, who understood that the definition of a boundary between air space and outer space at an altitude of 100 km above the Earth's surface is required for the definitive solution of a question that urgently needs a solution. He concluded that "no other criterion for demarcation would provide a more adequate solution".[47]

After reviewing the different proposals for delimitation of outer space, Robert F. A. Goedhart reached important conclusions, all of which are particularly relevant to the present study: "while awaiting the technological and economic advances of air travel and space travel, it can already be said with certainty that (1) the floor of outer space is identical with the lowest satellite orbits (i.e. the so-called perigee rule) by virtue of international custom (de lege lata), and that (2) it is advisable to conclude a treaty on a boundary situated at an altitude of 100 km perpendicular to the mean sea level (de lege ferenda)".[48]

Such reasoning is similar to that presented by Gennady M. Danilenko, who affirmed that "although the examination of the current international practice cannot lead to a legal definition of the precise line of the boundary between air space and outer space it indicates that there exists a general consensus among states in relation to the following two points: (1) The region at and above the line determined by the lowest perigees of satellites so far placed in orbit (100(+-10) km above sea level) is not subject to the sovereignty of underlying states and therefore is outer space. (2) The boundary between air space and outer space does not lie above that line".[49]

Accordingly, Andrezj Górbiel, who conducted relevant academic researches on this topic, confirmed being in favor of a boundary between air space and outer space set at 100 km of altitude from sea level,[50] in accordance with a position paper presented by the author on behalf of the Polish delegation before the UNCOPUOS legal subcommittee in its 1979 session.[51]

Considering the recent progress of sub-orbital flights and the growing interest in private space activities, Frans G. von der Dunk argued that "it is time to seriously

---

[46] Carl Q. Christol. *Modern International Law of Outer Space*. New York, USA: Pergamon Press, 1982. p. 528.

[47] I. H. Ph. Diedericks-Verschoor. *An Introduction to Space Law*. 2. ed. Dordrecht, the Netherlands: Kluwer Academic Publishers, 1999. p. 21.

[48] Robert F. A. Goedhart. *The Never Ending Dispute: Delimitation of Air Space and Outer Space*. Paris, France: Frontières, 1996. p. 51.

[49] Gennady M. Danilenko. "The Boundary Between Air Space and Outer Space in Modern International Law: Delimitation on the Basis of Customary Law". *Proceedings of the Twenty-Sixth Colloquium on the Law of Outer Space*. IISL, Budapest, 1983. p. 74.

[50] Andrzej Górbiel. *Legal Definition of Outer Space*. Lodz, Poland: Uniwersytet Lódzki, 1980. p. 73/74.

[51] A/AC.105/C.2/SR.304.

reconsider whether we should not firmly but flexibly establish that the boundary between airspaces and outer space at an altitude of 100 km, following the considerable number of instances where this number has already been referred to. After all, what is wrong with a nice round figure?"[52]

The 100 km above mean sea level standard seems to be the best one available since, at this altitude, there is just a residual part of Earth's atmosphere, while still being possible for space objects to hold an orbital position around the planet. Therefore, from a scientific standpoint, such altitude seems reasonable and justified. The existence of air above this mark should not be taken for granted, but the capacity of such uttermost layers of atmosphere to provide aerodynamic lift is minimal, to say the least.[53] On the other hand, space objects allocated below the proposed standard would be subjected to so much drag that their trajectory would be fatally compromised, forcing reentry unless alternative mechanisms of propulsion were available.[54]

The choice of the 100 km of altitude standard is not arbitrary, since takes into consideration the natural properties of upper atmospheric layers. Anyway, as observed by Marco G. Marcoff, scientific considerations will not be definitive for the proposal of such limit, since there is no clear natural boundary between air space and outer space, but rather a subtle and complex transition.[55] Maureen Williams posited that the spatial approach "should be based on a legal criterion rather than on technical, biological or other factors".[56] Manfred Lachs also manifested himself towards a purely conventionally established boundary, only incidentally linked to a specific environment or to a functional element.[57]

The delineation of horizontal national frontiers through air space, up to the proposed 100 km limit hereby proposed, shall be done using conic projection, in order to avoid the existence of empty spaces, which in fact would be created if a parallel

[52] "(…) it is time to seriously reconsider whether we should not firmly but flexibly establish that the boundary between airspaces and outer space at an altitude of 100 km, following the considerable number of instances where this number has already been referred to. After all, what is wrong with a nice round figure?" Frans G. Von der Dunk. "The Sky is the Limit – but Where does it End?" *Proceedings of the Forty-Eighth Colloquium on the Law of Outer Space*. IISL, Fukuoka, 2005. p. 92.

[53] As provided by the Oxford Dictionary of Space Exploration: "above 120 km, very little atmosphere remains, so objects can continue to move quickly without extra energy. The space between the planets is not entirely empty, but filled with tenuous gas of the solar wind as well as dust". E. Julius Dasch. *Oxford Dictionary of Space Exploration.* Oxford, England: Oxford University Press, 2005. p. 307.

[54] Roger R. Bate, Donald D. Mueller e Jerry E. White. *Fundamentals of Astrodynamics*. New York, USA: Dover, 1971. p. 152/159.

[55] Marco G. Marcoff. *Traité de Droit International Public de l'Espace*. Fribourg, Switzerland: Éditions Universitaires Fribourg, 1973. p. 323.

[56] Maureen Williams. "The Problem of Demarcation is Back in the Limelight". *Proceedings of the Twenty-Second Colloquium on the Law of Outer Space*. IISL, Munich, 1979. p. 248.

[57] "There is however the possibility of adopting a purely conventional boundary which would only subsidiarily rely on specific environmental or functional criteria." Manfred Lachs. *The Law of Outer Space*. Leiden, the Netherlands: Martinus Nijhoff, 2010. p. 56.

projection were to be applied.[58] By its turn, the vertical demarcation shall follow the natural specificities of our planet's surface, always taking into consideration the mean sea level as a standpoint. Indeed, the Earth is not a perfect sphere, but a pear shaped ovoid, with the Southern hemisphere slightly bigger than the Northern hemisphere and an equatorial zone with a radius superior to the polar one. Those natural features shall be reflected in the boundary between air space and outer space.

In his defense of a treaty that could prevent an arms race in space, Caesar Voûte argued that an arbitrary delimitation based at 100 km of altitude above mean sea level is utterly necessary, in order to distinguish different arms systems whose legality would differ if allocated in air space or outer space: "space weapon can be subdivided according to deployment mode into: ground-space; space-space; space-air; space-ground; air-space. We are dealing therefore with weapons either based in space aimed at targets in space, in the air, or on the ground, or with weapons aimed at targets in space, based on the ground or in the sea. These weapons can be further subdivided into nuclear and non-nuclear weapons. The deployment of some of these systems is banned by a number of treaties or Customary International Law. It is therefore important to delimitate air space, where their deployment would still be authorized in principle unless prohibited by other treaties, from outer space, where their deployment is not authorized."[59]

As far as international appraisal to the 100 km above mean sea level standard is concerned, Australia arguably contemplated it in its national legislation,[60] while several delegations before the UNCOPUOS made reference to it throughout the sessions, including, for instance, the former USSR.[61]

Nevertheless, one must acknowledge that the demarcation of the air space/outer space frontier, while guaranteeing the national security of territorial States, creates challenges to the use of outer space, noticeably as far as its access is concerned, due to the fact that national sovereignty would be applicable up to that final altitude. Indeed, as observed by Armand D. Roth, "the question is (…) more political than technical and the proliferation of theories on this subject seems to be inversely proportional to the perspective of solution."[62] Consequently, the provision of an international regime of passage rights in relation to the launching and reentry of space objects must be advanced.

---

[58] As clearly explained by Nicolas Mateesco Matte, who presented graphics to distinguish the parallel and conical projections. *Aerospace Law*. Toronto, Canada: Carswell, 1969. p. 46/47.

[59] Caesar Voûte. "Boundaries in Space", *In*: Bhupendra Jasani (ed.). *Peaceful and Non-Peaceful Uses of Space*. New York, USA: Taylor & Francis, 1991. p. 31.

[60] Lei de Atividades Espaciais, n. 34 1998-99, de 25 de novembro de 1998.

[61] A/AC.105/C.2/L.121, em 1979 e A/AC.105/C.2/L.139, em 1983.

[62] "La question est en réalité (…) plus politique que technique et la prolifération des théories en la matière semble inversement proportionnelle aux perspectives de solution". Armand D. Roth. *La Prohibition de l'Appropriation et les Régimes d'Accès aux Espaces Extra-Terrestres*. Paris, France: Presses Universitaires de France, 1992. p. 99.

Passage Rights

The intrinsic relationship between sovereignty/territory is addressed by International Law, recognizing the relevance of frontiers, which divide and draw near autonomous legal systems. Indeed, States are territorially organized entities that establish their mutual relationships over legal rules providing specific zones of competence, as indicated Bin Cheng, in such a way that Space Law cannot contradict the fundamental structure of the whole international legal system.[63]

The warning of Nicolas Mateesco Matte must not be ignored however: "if the national 'air space' were limited to 'navigable air space' (up to 30 or 40 km), there would be no difficulties as to the trajectories of space vehicles. But if the national 'air space' includes the 'non-navigable air space' (up to 80 km) or reach even higher, the problem of the exploration of space would become very complicated, since such an activity would require the approval of each subjacent state whose air space is utilized."[64]

Since the delimitation hereby proposed includes what Matte called "non-navigable air space", the concept of peaceful passage introduced by the USSR in its working papers to UNCOPUOS should be revaluated. As observed by Eilene Galloway, the Space Law treaties do not have rules exclusively linked to a certain territory, but also to specific activities, as is the case of the Liability Convention of 1972, which provides absolute liability for damages caused on the surface of the Earth or to aircraft in flight, irrespective of the success of the launching. Thus, Galloway concluded that to only draw the boundary between air space and outer space would not suffice, since other questions need discussion,[65] for instance the legal regime applicable to launching and reentry of space objects. The resort to passage rights for peaceful space activities may very well provide a long pursued solution.

In their study regarding passage rights and space activities, Marietta Benkö and Jurgen Gebhard introduced scientific and technical aspects that allow a better grasp of the problem. At launch, in order to compensate for the air drag that opposes the lift force of the rocket motor, the trajectory of the rocket is chosen to be very steep, of some 70° above the surface or more, in order to pass through the denser layers of atmosphere in the shortest possible time. When an altitude around 20 km is achieved, the flight direction is gradually changed to lower elevations, in order to attain an almost flat trajectory for orbital positions, only possible at a certain speed. Usually, the space object accomplishes the 100 km mark in a position not far away,

---

[63] "It may well be that they are all mistaken. But what is not believed to be feasible is to develop rules in a tiny corner of international law in a way which, however laudable in the abstract, is contrary to the basic framework of the international legal system". Bin Cheng. *Studies on International Space Law*. Oxford, England: Clarendon Pr., 1998. p. 444.

[64] Nicolas Mateesco Matte. *Aerospace Law*. Toronto, Canada: Carswell, 1969. p. 38.

[65] Eilene Galloway. "Legal Aspects of International Cooperation in Space: Area and Functional Concepts in Defining Outer Space". *Proceedings of the Twenty-Sixth Colloquium on the Law of Outer Space*. IISL, Budapest, 1983. p. 203.

horizontally speaking, of the launching center—probably just a few hundred kilometers from it. Therefore, at first sight, only those launchings taking place next to the borders of other States would represent relevant problems, as far as passage rights are concerned; but one must also take into consideration the different stages used by space rockets and their respective place of landing after launch.

The authors presented data regarding the American Saturn V rocket, whose second stage (weighting around 6 tons) impacted at surface distant between 4,000 and 5,000 km downrange. Due to that, preferred launching trajectories usually provide overflight of the territory of the respective Launching State or the high seas. As noted by Benkö and Gebhard, "serious problems might arise from the possible impact of spent rocket parts on foreign territory, if States which do not have suitable locations on their territories which satisfy the safety regulations normally imposed on launching sites should nevertheless wish to build their own launching facilities. In this case, the cooperation of any State potentially affected by such undertaking has to be ensured, since safety measures will have to be taken by this State, e.g., evacuation of certain areas or arrangements to clear the flight path below the space object from international aviation. Moreover, the foreign State concerned might wish that space trajectories in the range of which satellite parts can still impact on the Earth should not lead over populated areas."[66]

When considered the reentry of space objects, the question gains even further complexity. Benkö and Gebhard presented graphics indicating that such falling objects cross the 100 km standard over 2,000 km away from the expected touchdown area. Therefore, the air space of many States may be affected by the free fall trajectory, providing a relevant danger to air traffic. It must also be acknowledged that the number of space debris has increased dramatically throughout the years, and their reentry trajectory, usually uncontrolled, is quite frantic and unpredictable. Regarding space planes, the authors offered prevision of a distance around 8,000 km between the place of atmospheric reentry (at 100 km of altitude from sea level) and the landing point.[67]

Returning trajectories of space objects are particularly dangerous when the mark of 60 km of altitude is reached since, from then on, it would be advisable that territorial States closed their air space around the falling trajectory in order to avoid collisions with aircraft in flight, through a coordinated effort which should be performed in cooperation between the Launching State and the respective air traffic authority of the State overflown.

In 1992, the Russian delegation to the UNCOPUOS presented a working paper about aerospace objects which indicated an interesting practice: in relation to the launching of satellites, there is a common practice, based on mere courtesy, of

[66] Marietta Benkö and Kai-Uwe Schrogl (eds.). *International Space Law in the Making: Current Issues in the UN Committee on the Peaceful Uses of Outer Space*. Paris, France: Frontières, 1993. p. 116/117.

[67] Marietta Benkö and Kai-Uwe Schrogl (eds.). *International Space Law in the Making: Current Issues in the UN Committee on the Peaceful Uses of Outer Space*. Paris, France: Frontières, 1993. p. 119/121.

Launching States providing prior information to every nation whose air space may be crossed below an altitude of 100 km from mean sea level.[68] According to the Russian representatives, there would not be an obligation for the Launching State to provide a prior communication of its activities, nor protests were presented by those States whose air spaces were crossed during launchings. However, it shall be noticed that this practice, if constant, may evolve to a *consuetudo* regarding passage rights, which, if followed by eventual *opinio juris* among States, could develop into a new international custom.[69]

It is debatable that passage rights for launching and reentry of space objects have already achieved the level of consuetudinary international rule, at least up to this moment. States remain very concerned in reaffirming, whenever necessary, that their national sovereignty extends to their air space, as in fact is assured by International Law. Therefore, only through an international agreement passage rights could be effectively provided in Space Law, avoiding future international disputes.[70]

In this aspect, Tanja Masson-Zwaan defended the regulation of the right of free passage through foreign air space for space objects, possibly with cooperation with ICAO: "one might argue that such 'right of overflight' have never been contested since numerous satellites have been put into orbit without opposition by the 'overflown States'', and that this 'tacit consent' has resulted in acquiescence by the community of States. On the other hand, the relatively short period of time involved here may be an obstacle to the possible existence of a rule of customary international law. It is therefore still necessary to codify this rule in a legal instrument."[71]

In order to conceive proper rules regarding passage rights for space objects, the Space Treaty of 1967 must be taken into consideration, which established the right of every nation to freely access outer space in the second paragraph of its first article: "outer space, including the Moon and other celestial bodies, is not subject to

---

[68] A/AC.105/C.2/L.189.

[69] Marietta Benkö and Kai-Uwe Schrogl (eds.). *International Space Law in the Making: Current Issues in the UN Committee on the Peaceful Uses of Outer Space*. Paris, France: Frontières, 1993. p. 130.

[70] In opposite direction, Jiri Malenovsky understood that there already exists an international custom regarding passage rights from 40/45 km of altitude above sea level: "this zone (…) is obviously free for passage to and from outer space, not subject to unilateral limitations and licensing regimes by underlying states". "Right of Passage into Outer Space". *Proceedings of the Thirty-Third Colloquium on the Law of Outer Space*. IISL, Dresden, 1990. p. 325. Same interpretation was once supported by Manfred Lachs: "On many occasions the [space] object moved through airspace of the launching state, but it may have travelled over other lands. These were never asked for permissions to cross their airspace and never protested against the journey through it. Thus one might assume that a customary law has developed: of innocent passage into outer space. This practice continues." "Freedoms of the Air – the Way to Outer Space", *In:* Tanja Masson-Zwaan e Pablo Mendes De Leon. *Air and Space Law: De Lege Ferenda – Essays in Honour of Henri A. Wessenbergh*. Dordrecht, the Netherlands: Martinus Nijhoff, 1992. p. 244.

[71] Tanja Masson-Zwaan. "The Influence of the Achievements and Failures of the Past on the Future of Outer Space Law", *In:* Tanja Masson-Zwaan (ed.). *Space Law: Views of the Future*. Deventer, the Netherlands: Kluwer, 1988. p. 41.

national appropriation by claim of sovereignty, by means of use or occupation, or by any other means." Hence, there is a reasonable justification for providing a body of internationally binding rules granting passage rights for launching and reentry of space objects, in order to safeguard the right of every State, irrespective of the size of their territories or their location, to develop their own launching centers.

The regulation of passage rights must also reflect Article Three of the Space Treaty:

> States Parties to the Treaty shall carry on activities in the exploration and use of outer space, including the Moon and other celestial bodies, in accordance with international law, including the Charter of the United Nations, in the interest of maintaining international peace and security and promoting international cooperation and understanding.

Therefore, the passage of space objects through foreign air space cannot be considered peaceful if conducted in violation of International Law, if disrespecting the sovereign rights of the territorial State or if implicating unjustified dangers to its national security or the wellbeing of the local population or environment. Consequently, as far as unpeaceful passages are concerned, the territorial State shall have the right to take proportionate measures of self-defense.

It must be recalled that such reasoning will not be fully applicable if the functionalist approach is considered, since it connects the legal regime to the purpose of the activity, irrespective of where it takes place. In accordance to the functional point of view, a space object, during launching or reentry, would always follow Space Law, not being subjected to national sovereignty claims. What constitutes a space object would in fact lie in the hands of the appropriate Launching State, irrespective of the national interests of the eventual territorial States.

The concept of peaceful passage, hereby supported, must be understood as meaning non-aggressive, that is, "not constituting an attack upon, or a threat against, the territorial integrity and independence of another State", as underlined by Robert F. A. Goedhart, since arguably that is the prevailing interpretation of the Space Treaty. Nowadays, it is commonly understood that the use of military space satellites, as long as not allocating nuclear weapons or weapons of mass destruction, and not designed for aggressive endeavors, are in accordance to International Law. As a matter of fact, one of the most important branches of outer space activities is the one concerned with spy satellites, which were contemplated theoretically even before the technology to put them in orbit had been made available.[72]

Although we are dealing with a very specific reality, it seems reasonable to consult selected provisions of the Montego Bay Convention on Law of the Sea of 1982, in order to better comprehend passage rights in an important International Law context, that is, the innocent and transit passage of foreign ships through territorial sea.[73] As explained by Donald R. Rothwell and Tim Stephens, "there is a

---

[72] Robert F. A. Goedhart. *The Never Ending Dispute: Delimitation of Air Space and Outer Space.* Paris, France: Frontières, 1996. p. 15.

[73] Robert F. A. Goedhart. *The Never Ending Dispute: Delimitation of Air Space and Outer Space.* Paris, France: Frontières, 1996. p. 15/17.

strong sense that the counter-balancing of a 12 nm [nautical miles] territorial sea with rights of innocent and transit passage was an essential part of the 'package deal' of the LOSC [1982 UN Convention on the Law of the Seas] which reflected in this particular instance the differing views of coastal and maritime states. (...) Notwithstanding the navigational rights granted to foreign shipping within a coastal state's territorial sea, the coastal state nevertheless retains significant rights so as to maintain its sovereignty and national security. It has a capacity to apply and enforce its laws against foreign merchant shipping, and has a limited right to temporarily suspend innocent passage."[74]

The study of the Montego Bay Convention provisions regarding passage rights through the territorial sea is advised since, as concluded by Hasjim Djalal, the freedom of the sea has become a highly regulated area.[75] Indeed, article 18 of the Montego Bay Convention defines "passage" as a continuous and expeditious navigation through the territorial sea, without entering internal waters or scales,[76] while the next article states that shall be considered innocent the passage those which are "not prejudicial to the peace, good order or security of the coastal State. Such passage shall take place in conformity with this Convention and with other rules of international law".

Accordingly, in its article 19, said treaty established that a passage will not be considered innocent where any of the following activities is performed:

(a) any threat or use of force against the sovereignty, territorial integrity or political independence of the coastal State, or in any other manner in violation of the principles of international law embodied in the Charter of the United Nations;
(b) any exercise or practice with weapons of any kind;
(c) any act aimed at collecting information to the prejudice of the defence or security of the coastal State;
(d) any act of propaganda aimed at affecting the defence or security of the coastal State;
(e) the launching, landing or taking on board of any aircraft;
(f) the launching, landing or taking on board of any military device;
(g) the loading or unloading of any commodity, currency or person contrary to the customs, fiscal, immigration or sanitary laws and regulations of the coastal State;
(h) any act of willful and serious pollution contrary to this Convention;

---

[74] Donald R. Rothwell and Tim Stephens. *The International Law of the Sea.* Oxford, England: Hart Publishing, 2010. p. 76.

[75] "In my mind it would be very difficult now to say that the freedom of the sea is still as before. I would even say that the freedom of the sea has become a highly regulated sea. From all sides and all kinds of views there is a highly regulated use." Hasjim Djalal. "Remarks on the Concept of 'Freedom of the Seas", *In:* Myron H. Nordquist, Tommy T. B. Koh and John Norton Moore (ed.). *Freedom of Seas, Passage Rights and the 1982 Law of the Sea Convention.* Leiden, the Netherlands: Martinus Nijhoff, 2009. p. 69.

[76] "Article 18. Meaning of passage. 1. Passage means navigation through the territorial sea for the purpose of: (a) traversing that sea without entering internal waters or calling at a roadstead or port facility outside internal waters; or (b) proceeding to or from internal waters or a call at such roadstead or port facility. 2. Passage shall be continuous and expeditious. However, passage includes stopping and anchoring, but only in so far as the same are incidental to ordinary navigation or are rendered necessary by force majeure or distress or for the purpose of rendering assistance to persons, ships or aircraft in danger or distress."

(i) any fishing activities;
(j) the carrying out of research or survey activities;
(k) any act aimed at interfering with any systems of communication or any other facilities or installations of the coastal State;
(l) any other activity not having a direct bearing on passage.[77]

Observed E. D. Brown that all those activities are very clearly of such a nature as to render the passage of a foreign vessel as non-innocent. But it must be noticed that those provisions "make 'innocence' depend on conduct during passage and prevents the coastal State from determining the non-innocence of a passage by reference to the character of the vessel, quite independent of its conduct during passage."[78]

With the needed adaptations, those Law of Sea's rules could be applied to Space Law, since they present important principles for the regulation of passage rights. Indeed, similar problems may benefit from similar solutions, based on an analogical approach.[79]

Besides, as observed by Robert F. A. Goedhart, the Montego Bay Convention in its articles 124 to 132 (Part X) provides that land-locked States have free access to and from the sea through the territory of transit States.[80] In fact, as indicated by Kishor Uprety, "it may be argued that under certain conditions, the grant of transit freedom for LLS [landlocked States] is an obligation of the State of passage, independent of all international agreements. Freedom of transit is thus not a 'right' that any State can exercise in other transit States without prior consent. In order to be eligible to claim this right, the demanding State must fulfill certain eligibility criteria."[81]

The same fate is endured by small States, especially those with multiple nearby sovereign neighbors, as far as launching of space objects is concerned. To J. N. Singh, the right of access to outer space, provided by article first, paragraph second of the Space Treaty, could only be effective if passage rights over the air space of other States were recognized during launching.[82]

---

[77] "Article 19. Meaning of innocent passage. 1. Passage is innocent so long as it is not prejudicial to the peace, good order or security of the coastal State. Such passage shall take place in conformity with this Convention and with other rules of international law."

[78] E. D. Brown. *The International Law of the Sea.* Vol. I. Aldershot, USA: Dartmouth, 1994. p. 57.

[79] It important to recall that the coastal State may provide maritime routes for foreign vessels during their innocent passage through the territorial sea: "O Estado costeiro designará rotas marítimas e prescreverá sistemas de separação de tráfego, quando necessário à segurança da navegação, exigindo que navios estrangeiros os utilizem, principalmente, navios de propulsão nuclear e similares, a quando da passagem inocente, levando em conta as características dos navios e dos canais utilizados para a navegação internacional e a densidade de tráfego." Adherbal Meira Mattos. *O Novo Direito do Mar.* 2. ed. Rio de Janeiro, Brazil: Renovar, 2008. p. 21.

[80] Robert F. A. Goedhart. *The Never Ending Dispute: Delimitation of Air Space and Outer Space.* Paris, France: Frontières, 1996. p. 17.

[81] Kishor Uprety. *The Transit Regime for Landlocked States.* Washington D.C., USA: the World Bank, 2006. p. 29.

[82] J. N. Singh. *Outer Space, Outer Sea, Outer Land and International Law.* New Deli, India: Harnam Publications, 1987. p. 38.

The International Court of Justice already decided in favor of passage rights through territorial sea, including straits, as in the Corfu Channel Case of 1948, between the UK and Albania,[83] as well as by land, when supported by local customs or special circumstances, as faced at the analysis of a Portuguese claim against India in 1960.[84] Consequently, there is sufficient international legal jurisprudence to support the provision of passage rights for space activities, regarding launching and reentries of space objects, as long as they are peaceful, do respect International Law and do not pose relevant dangers to the territorial State.

Passage rights should only be applicable if deemed peaceful, as established by international standards, respecting the safety of the territorial State, which, otherwise, could be entitled to promote protective alternatives to safeguard its national interests in extreme circumstances. Accordingly, it is hereby supported that the passage of space objects through foreign air space should not be considered peaceful if conducted in violation of International Law, disrespecting the sovereign rights of the territorial State or implicating unjustified dangers to its national security or the wellbeing of the local population and environment.

By all accounts, any regulation of space objects' passage rights must take into consideration the interests of the territorial State, especially in relation to its national security. Up to this moment, one must acknowledge the fact that States may occasionally face the inconvenient reality of having their air space crossed by space objects on their way up, without enjoying proper support by Space Law, as far as their territorial privileges are concerned.

Therefore, the State whose air space is traversed by a foreign space object should have the right to receive prior information of those activities, in order to prepare any security measures deemed necessary, including the closure of its air space for commercial flights. Routes for passage of foreign space objects during launching, whenever provided by the relevant territorial State, should also be considered and respected. On the other hand, the territorial State must not be allowed to demand

---

[83] "In this case, British warships passing through the straits were fired upon by Albanian guns. Several months later, an augmented force of cruisers and destroyers sailed through the North Corfu Channel and two of them were badly damaged after striking mines. This impelled the British authorities to sweep the Channel three weeks later, and to clear it of some twenty mines of German manufacture. The Court, in a much-quoted passage, emphasized that: 'states in time of peace have a right to send their warships through straits used for international navigation between two parts of the high seas without the previous authorization of a coastal state, provided that the passage is innocent." Malcolm N. Shaw. *International Law*. 3. ed. Cambridge, England: Cambridge University Press, 2003. p. 513.

[84] "There are forty-one land-locked states and principalities in existence and numerous enclaves detached from a parent entity (and lacking access to the sea). Rights of transit, particularly for trade purposes, are normally arranged by treaties, but they may exist by revocable license or local custom. A right of transit may be posited as a general principle of law itself or on the basis of a principle of servitudes or other general principles of law. However, a general right of transit is difficult to sustain, and the principle of servitudes, and the other possibly available instruments, are controversial and depend, in any case, on the existence of special circumstances." Ian Brownlie. *Principles of Public International Law*. 6. ed. Oxford, England: Oxford University Press, 2003. p. 271.

unreasonable requirements or charges for the crossing of its air space by foreign space objects, which could otherwise implicate in an indirect denial of international access to outer space.

One must recognize the challenges imposed by reentries on any regulation about passage rights of space objects, since, most of the times, due to technical imperatives, no control is maintained by the respective Launching State.[85] Nevertheless, even in those circumstances, the territorial State shall have the right to take emergency actions regarding its national air space, hereby proposed as extending up to 100 km of altitude above mean sea level, including the apprehension and destruction of the falling space object representing clear and immediate danger to people, property or the environment, in accordance with a *de lege ferenda* international regulation exempting responsibility for such last resort maneuvers.

Currently, there is no treaty provision regarding the possible defensive actions that the territorial State may take in relation to uncontrolled space debris already traversing its air space. Many inoperative satellites, currently in orbit, have considerable mass and surface, enough for at least components or parts thereof to survive reentry. Some of them carry dangerous cargo, including toxic fuels and nuclear reactors.

Justification for emergency actions against free falling space objects, considering the current international regulation, could only be advocated based on factors that exclude responsibility,[86] for instance, a state of necessity or legitimate defense, mostly consuetudinary concepts based on general principles of Law subjected, therefore, to eventual conflicting interpretation.[87]

While reviewing the legality of destruction of space objects that represent danger to the security of a certain State, not due to its activities *per se*, but because they are falling out of control and with a high chance of surviving reentry, Bruce A. Hurwitz

---

[85] Marietta Benkö and Kai-Uwe Schrogl (eds.). *International Space Law in the Making: Current Issues in the UN Committee on the Peaceful Uses of Outer Space*. Paris, France: Frontières, 1993. p. 116/121.

[86] "A doutrina e a prática internacionais têm geralmente admitido que, em certos casos, devido a circunstâncias especiais, a responsabilidade do estado desaparece. Tais casos são: 1°. Aqueles em que o ato perde o caráter ilícito, transformando-se no exercício de um direito reconhecido; 2°. Aqueles em que o ato determinante da responsabilidade, apesar de ilícito em si mesmo, não pode acarretar as consequências naturais dos fatos ilícitos; 3°. Aqueles em que o decurso do tempo extingue a responsabilidade; 4°. Aqueles que representam a consequência direta do comportamento inconveniente e censurável do indivíduo lesado." Hildebrando Accioly, G. E. do Nascimento e Silva and Paulo Borba Casella. *Manual de Direito Internacional Público*. 16. ed. São Paulo, Brazil: Saraiva, 2008. p. 362.

[87] It is important to recall that article 51 of the United Nations Charter, of 1945, presents a general provision regarding legitimate defense: "nothing in the present Charter shall impair the inherent right of individual or collective self-defence if an armed attack occurs against a Member of the United Nations, until the Security Council has taken measures necessary to maintain international peace and security. Measures taken by Members in the exercise of this right of self-defence shall be immediately reported to the Security Council and shall not in any way affect the authority and responsibility of the Security Council under the present Charter to take at any time such action as it deems necessary in order to maintain or restore international peace and security".

claimed that those defensive measures could only be considered legal under the condition that "such destruction does not itself result in more interference or in environmental contamination, and that it is carried out as a last resort."[88]

During the most famous case of space debris collision on the surface of a State other than the one of launching, Canada complained that the Soviet satellite "Cosmos 954" had violated its air space during reentry. In fact, soon after the fall of the space object, in 1978, the Canadian government presented a diplomatic claim before the USSR based not only on the environmental damages caused by the satellite's nuclear reactor, but also due to breach of national sovereignty produced by its unauthorized reentry through national air space. Canada considered that its national borders were breached without prior authorization, which indeed was not specifically denied by the Soviets.[89] In the end, both parties agreed on a compromised solution, by which the USSR accepted to pay compensation in a certain amount, considerably lower than the one originally requested by Canada, but gave away its right to recover the surviving parts of its satellite.[90]

Additionally, the evolution of technology may, in due time, require further discussion of the problem of passage rights, as affirmed by Marietta Benkö and Engelbert Plescher, "when new reusable space transportation systems and mobile launch systems, which are already in planning phase, should become operational."[91]

In conclusion, it seems reasonable to consider that a passage of space object through foreign air space shall be considered peaceful if performed in accordance to International Law, without posing reasonable risks to the territorial State. By its turn, the territorial State in question should not be allowed to demand charges for the crossing of its air space. However they should have the right to receive prior information of those activities, in order to prepare any security measures deemed necessary, including the closure of its air space for commercial flights to avoid any risk of collision with aircraft in flight. Routes for passage of foreign space objects during launching prepared by the relevant territorial State should be respected as well, for the same reasons already stated.

[88] Bruce A. Hurwitz. *The Legality of Space Militarization*. Amsterdam, the Netherlands: North-Holland, 1986. p. 151.

[89] "Under section (b) of the Claim, 'General Principles of International Law', the Canadians established that '[t]he intrusion of the … satellite into Canada's air space and the deposit on Canadian territory of hazardous radioactive debris from the satellite constitutes a violation of Canada's sovereignty.' Specifically, the intrusion and damage represented 'interference with the sovereign right of Canada to determine the acts that will be performed on its territory. International precedents recognize that a violation of sovereignty gives rise to an obligation to pay compensation." Bruce A. Hurwitz. *State Liability for Outer Space Activities: in Accordance with the 1972 Convention on International Liability for Damage Caused by Space Objects*. Dordrecht, the Netherlands: Kluwer Academic Publishers, 1992. p. 120.

[90] Göran Lysén. *State Responsibility and International Liability of States for Lawful Acts: a Discussion of Principles*. Gothenburg, Sweden: Iustus Förlag, 1997. p. 141.

[91] Marietta Benkö and Engelbert Plescher. *Space Law – Reconsidering the Definition/Delimitation Question and the Passage of Spacecraft Through Foreign Airspace*. The Hague, The Netherlands: Eleven, 2013. p. 48.

Those rules cannot be effectively applicable for reentry of space objects, which most of the times cannot be controlled by the Launching States. Nevertheless, in those cases, the territorial State shall have the right to take emergency actions up to the limits of its air space, hereby proposed as extending up to 100 km of altitude above mean sea level, including the apprehension and destruction of the falling space object, following a future regulation which excludes international State responsibility for such last resort maneuvers.

In order to clearly solve this relevant situation, which affects the international community as a whole, a written and carefully drafted rule is favored, to be encompassed in a new treaty or protocol to Air Law or Space Law conventions. Otherwise, doubts would remain in relation to any possible custom that could, in one way or another, authorize passage rights for space activities. As a matter of fact, Robert F. A. Goedhart concluded that:

> At any rate, the passage of spacecraft through the air space of other States without prior consent, whether deliberate or resulting from miscalculation or misfire, would apparently constitute a violation of the territorial sovereignty of those States. The only exception to the rule would seem to be circumstances beyond the control of the launching State, if proved, upon a spacecraft's reentry into the atmosphere. Planned reentry into the atmosphere and landing on the territory of another State without prior consent would thus not qualify as an exception in the sense just indicated.[92]

Without an international compromise which conciliates both spatialists and functionalists principles, the tendency is for the question to be unilaterally considered by States through their national legislation, contributing for an uncertain scenario where conflicts of interpretation may impair the free uses of outer space.[93]

Even Gyúla Gál, internationally recognized as one of the most important supporters of the functional approach, once admitted that a "compromise proposal" would constitute an important attempt to harmonize those opposite concepts, and concluded: "to my humble opinion an international agreement following this line would successfully end the 'never ending dispute'."[94]

It is thus essential to clearly provide a vertical limit to State sovereignty by treaty at the standard of 100 km of altitude from sea level, beyond which shall lie the outer space, at the same time establishing passage rights applicable for launching and reentry of space objects through appropriate international regulation respecting the sovereign rights of the territorial State.

This is the best solution of a still unsolved International Law problem, which claims serious consideration, among scholars and diplomats, after decades of fruitless debates. Viable alternatives must be contemplated to address such a relevant global concern. Time is ticking, and no debate shall last forever.

---

[92] Robert F. A. Goedhart. *The Never Ending Dispute: Delimitation of Air Space and Outer Space*. Paris, France: Frontières, 1996. p. 20.

[93] Alexandra Harris; Ray Harris. "The Need for Air Space and Outer Space Demarcation". *Space Policy*. Vol. 22, n. 1. Dayton, USA, Fevereiro 2006. p. 3/7. E. G. Vassilevskaya. "Delimitation of Air and Space – Major Issue for Ensuring Peace and Security". *Proceedings of the Twenty-Eighth Colloquium on the Law of Outer Space*. IISL, Stockholm, 1985. p. 114/117.

[94] Gyúla Gál. "Thirty Years of Functionalism". *Proceedings of the Fortieth Colloquium on the Law of Outer Space*. IISL, Turim, 1997. p. 130.

Chapter 8
Final Remarks

> Nature had given all things to all men. Nevertheless, owing to the fact, that the distances separating different regions prevent men from using many of goods desirable for human life (since not all things are produced in all localities (…)), passage to and fro was found to be a necessity. [Hugo Grotius (1583–1645)] Mare Liberum[1]

A clear delimitation of the frontier between air space and outer space is hereby advocated, through a compromise that contemplates features of both the functionalist and spatialist approaches in a single and comprehensive legal regime, assuring legal security for activities conducted in both realms.

The relevance of this topic can be observed at the UNCOPUOS, where discussions regarding the delimitation and definition of outer space have recovered relevance, due to a scientific and technical progress that, each day, narrows the distance between aeronautic and astronautic activities. Questions related to aerospace objects, sub-orbital flights, tether satellites and space tourism represent, therefore, challenges to the current international legal order, justifying a careful review of proposals for delimitation of the so-called "final frontier."

In 2011, during the historic 50th session of UNCOPUOS' legal subcommittee, the International Institute of Space Law (IISL) and the European Centre for Space Law (ECSL) sponsored the workshop "A New Look on the Delimitation of Airspace and Outer Space", where illustrious scholars presented relevant contributions to this subject matter in a common effort to assist international delegations in their diplomatic activities.[2] The workshop was presided by Tanja Masson-Zwaan and Sergio Marchisio, respective presidents of IISL and ECSL. It is possible to ascertain a clear feature in all presentations, that is, a pleading for a better appreciation of legal aspects involved with the delimitation and definition of outer space, in a distinctive

[1] Hugo Grotius. *Mare Liberum.* Leiden, the Netherlands: Brill, 2009. p. 127.

[2] http://www.oosa.unvienna.org/oosa/en/COPUOS/Legal/2011/symposium.html, accessed on 04.11.2011.

© Springer International Publishing Switzerland 2015

O.O. Bittencourt Neto, *Defining the Limits of Outer Space for Regulatory Purposes*, SpringerBriefs in Space Development,
DOI 10.1007/978-3-319-16685-8_8

and comprehensive way, taking into consideration both the scientific evolution and the need of safeguarding international legal security.[3]

I was granted the honor to act as rapporteur of this illustrious event, which covered presentations from Catherina Doldirina, of McGill University[4]; Luboš Perek, of the Learned Society of the Czech Republic[5]; Marco Pedrazzi, from the University of Milan[6]; Joanne Irene Gabrynowicz, from the University of Mississippi[7]; Sang-Myon Rhee, of the University of Seoul[8]; and finally Jean-François Mayence, from Belgium "Politique Scientifique Fédérale".[9]

During the sessions of the legal subcommittee in that year, arguably due to the influence of said workshop, several delegations presented formal opinions favorable to the delimitation of the frontier between air space and outer space,[10] including Ukraine, Brazil, Russia, Indonesia, Venezuela, Algeria, Chile, Czech Republic, Saudi Arabia, Morocco and even the Netherlands, which, contradicting its prior position on this matter, affirmed that eventually a national delimitation through municipal legislation may prove to be needed, considering the perspectives of use of Dutch territory as platform for commercial suborbital flights.[11] On the other hand, the USA maintained its traditional understanding that any attempted delimitation would only constitute a theoretical exercise, incapable of foreseeing the effects of scientific progress to space activities.

Considering the declarations presented by the diplomatic delegations before the UNCOPUOS throughout the years, it is possible to affirm that the voices favorable to delimitation are in a greater number than the ones against it.[12] However, since the legal subcommittee's decisions depend on the consensus among member States, a single opposition is enough to obstruct any resolution or treaty proposing a definitive solution to the matter at hand.

Nonetheless, from the outside looking in, one may see that the international concern about the delimitation of the vertical frontier to State sovereignty has regained momentum. This is particularly true in view of new technological and commercial

---

[3] http://www.oosa.unvienna.org/pdf/limited/c2/AC105_C2_DEF_2011_L01E.pdf, accessed on 04.14.2011.

[4] http://www.oosa.unvienna.org/pdf/pres/lsc2011/symp01.pdf, accessed on 04.14.2011.

[5] http://www.oosa.unvienna.org/pdf/pres/lsc2011/symp02.pdf, accessed on 04.14.2011.

[6] http://www.oosa.unvienna.org/pdf/pres/lsc2011/symp03.pdf, accessed on 04.14.2011.

[7] http://www.oosa.unvienna.org/pdf/pres/lsc2011/symp04.pdf, accessed on 04.14.2011.

[8] http://www.oosa.unvienna.org/pdf/pres/lsc2011/symp05.pdf, accessed on 04.14.2011.

[9] http://www.oosa.unvienna.org/pdf/pres/lsc2011/symp06.pdf, accessed on 04.14.2011.

[10] Official report of the 50th Session of the UNCOPUOS legal subcommittee: http://www.oosa.unvienna.org/oosa/en/COPUOS/Legal/index.html. For unedited transcription: http://www.oosa.unvienna.org/oosa/en/COPUOS/Legal/transcripts/index.html.

[11] The former Dutch position, presented on 03.03.2010, defended that there was no need to provide the delimitation of outer space, but only of the terms "space activities" and "space objects", in a clear "functionalist" rationale. A/AC.105/865/Add.8.

[12] "Draft report of the Chair of the Working Group on the Definition and Delimitation of Outer Space", 50th Session, UNCOPUOS Legal Subcommittee: http://www.oosa.unvienna.org/pdf/limited/c2/AC105_C2_DEF_2011_L01E.pdf, accessed on 04.14.2011.

developments in aeronautic and space activities which are capable of modifying traditional positions, as demonstrated by answers offered by States to the UNCOPUOS questionnaire regarding the topic.[13] In fact, the position presented by the United Kingdom in 2010 must be duly acknowledged: "we anticipate that the development of space transportation systems functioning seamlessly between airspace and outer space, relying on lift to fly through the air for part of their flight profile, will create uncertainties about the legal regime applicable to them. In particular, the distinct liability regimes applicable to each may be conflicting. The United Kingdom is currently reviewing its licensing process and how it could relate to commercial human spaceflight, where this will likely be an issue. We recognize the need to avoid hybrid solutions and will seek a regulatory solution that provides seamless consideration and a degree of legal certainty for operators. (…) Although the United Kingdom is not considering the possibility of defining a lower limit of outer space and/or an upper limit of airspace, the United Kingdom may consider the possibility of enacting special international or national legislation relating to a mission carried out by an object in both airspace and outer space."[14]

It should be stressed that Brazil remains a strong advocate of the delimitation of outer space, having supported a coherent position ever since the first sessions of the legal subcommittee, through statements that emphasize the intrinsic differences between the legal regimes applicable to air space and outer space. In its 2009 formal reply to the questions proposed by the legal subcommittee on this regard, the Brazilian position was laid upon three major considerations: "1. The speed with which technological advances in space and aviation research are being made indicate that in the near future it will be possible to develop spacecraft with characteristics similar to those of an "aerospace object", which could be defined as an object capable of flying and performing activities both in outer space and in airspace. 2. Taking that into account, aerospace objects should be regulated by international space law when in outer space and by international and national air law when in airspace. The main distinction between those two legal regimes is that in air law the principle of State sovereignty prevails while in space law it does not. 3. In order to adequately deal with situations arising from the development or utilization of aerospace objects (for example, activities in foreign airspace), it is necessary for the international community to take measures to establish universally accepted principles and parameters leading to the definition of boundaries between outer space and airspace."[15]

---

[13] Up to this moment, presented formal answers to the refereed questionary the following States: the Netherlands, Tunisia, Denmark, Jordan, Austria, El Salvador, Algeria, Norway, United Kingdom, Mauritius, Czech Republic, Estonia, Bangladesh, Germany, Iraq, Serbia, Thailand, Azerbaijan, Qatar, Saudi Arabia, Belarus, Mexico, Brazil, Nicaragua, Ukraine, Iceland, Nigeria and Venezuela. http://www.oosa.unvienna.org/oosa/SpaceLaw/national/def-delim/question.html, accessed on 11.21.2011.

[14] A/AC.105/889/Add.8.

[15] A/AC.105/889/Add.2.

The absence of a proper multilateral solution consequently authorizes unilateral efforts of delimitation through municipal legislation. The incentives for regulation of national space activities are abundant, and found in several conventions. For instance, in accordance to the Space Treaty of 1967, States are internationally responsible for every local endeavor regarding exploration and use of outer space, whether performed by governmental or non-governmental entities (Article VI).

Therefore, space faring nations should be encouraged to develop national space legislation, especially whenever activities such as space tourism and suborbital flights are conducted from their territory. The above-mentioned Dutch declaration exemplifies the national concern that justifies the provision of a clear territorial jurisdiction. Based on recent efforts, one may infer a very realistic scenario in which the vertical limit of State sovereignty could become increasingly subject to different municipal provisions, solely supported by local, strategic interests. Suddenly, we may find ourselves in a complex scenario where the border of air space and outer space differs from one country to another, presenting amazing challenges to international cooperation and hindering the progress of space activities. If the Law of the Seas serves as an indicator, as far as national limits of the territorial sea are concerned, universal delimitation should always prevail upon local and perhaps egoistic solutions.

Taking into consideration this labyrinthine scenario, the Brazilian delegate José Monserrat Filho, president of the UNCOPUOS legal subcommittee study group on definition and delimitation of outer space, proposed in 2011 the composition of a group of international experts with the incumbency of preparing a proposal for solution of the current standoff during. Such an endeavor should be based on a compromising proposal between the functionalist and spatialist approaches. Since the initiative was accepted by the delegations, it remains to be seen if such alternative path may lead to a fruitful resolution.

The study presented herein intends to provide food for thought for studies on the definition and delimitation of outer space, offering an alternative that not only pays respect to past proposals but also includes particular conciliatory features.

First and foremost, the necessity is recognized of providing a clear legal standard of altitude for delimitation of the frontier between air space, subjected to exclusive State sovereignty, and outer space, that rests free from sovereign appropriation and open for international use and exploration.

This position is justified in order to safeguard one of the most important Space Law principles, that is, the prohibition of national claims over outer space, including orbital positions and celestial bodies.

The border between air space and outer space should be based on an arbitrary limit, established by an international instrument (favorably a treaty), in order to assure legal security for aeronautic and space activities and to avoid international disputes. The standard of 100 km of altitude from mean sea level is hereby advocated, indeed a mark that has been suggested by several publicists[16] and delegations

[16] For instance: I. H. Ph. Diedericks-Verschoor. *An Introduction to Space Law*. 2. ed. Dordrecht, the Netherlands: Kluwer Academic Publishers, 1999. p. 21; Robert F. A. Goedhart. *The Never Ending*

before the UNCOPUOS[17] and the Conference on Disarmament,[18] since it represents a singular zone where aerodynamic lift decreases to critical levels and where the lowest perigees attainable by space objects in orbit can reasonably be identified.[19]

In addition to those features that follow a spatialist logic, is also proposed the international regulation of passage rights for space objects, applicable during launchings and reentries irrespective of the success of their missions. Such element is, in contrast, connected to the functionalist rationale, since it applies a distinctive legal regime for space activities, whether carried out in air space or in outer space.

It must be reckoned that, for space objects to reach orbit and return from it, national or foreign air space ought to be overflown, generating sensitive legal and political issues.[20]

In order to conceive proper rules regarding passage rights for space objects, one should take special consideration to the Space Treaty, of 1967, that established the right of every nation regarding free access to outer space. Hence, there is a reasonable justification for providing a body of internationally binding rules that grant passage rights for space objects, in order to safeguard the prerogative of any State, irrespective to the size of their territories or their geographical location, of developing their own launching capabilities.

But passage rights should only be applicable if such transit were to be deemed peaceful, as established by international standards, respecting the safety of the

Dispute: Delimitation of Air Space and Outer Space. Paris, France: Frontières, 1996. p. 51; Gennady M. Danilenko. "The Boundary Between Air Space and Outer Space in Modern International Law: Delimitation on the Basis of Customary Law". *Proceedings of the Twenty-Sixth Colloquium on the Law of Outer Space*. IISL, Budapest, 1983. p. 74; Andrzej Górbiel. *Legal Definition of Outer Space*. Lodz, Poland: Uniwersytet Lódzki, 1980. p. 73/74.

[17] It is important to mention the proposals presented by the USSR: A/AC.105/C.2/L.121, 1979 and A/AC.105/C.2/L.139, 1983. For a comprehensive review of the current position of the Russian Federation on that matter: A/AC.105/889/Add.10.

[18] On 12 February 2008, Russia submitted to the Conference on Disarmament the "Draft Treaty on the Prevention of the Placement of Weapons in Outer Space, the Threat or Use of Force against Outer Space Objects", prepared jointly with China, defining "outer space" as "space beyond the elevation of approximately 100 km above ocean level of the Earth." For the latest discussion on that regard, reference is made to the thematic discussion held at the Conference on Disarmament on 31 July 2012: http://www.unog.ch/80256EDD006B9C2E/(httpNewsByYear_en)/23B0ED99B9B9B43BC1257A4C003F86B9?OpenDocument, accessed on 11.02.2012.

[19] Considering the recent progress of suborbital flights and the growing interest in private space activities, Frans G. Von Der Dunk defended that "it is time to seriously reconsider whether we should not firmly but flexibly establish that the boundary between airspaces and outer space at an altitude of 100 km, following the considerable number of instances where this number has already been referred to. After all, what is wrong with a nice round figure?" "The Sky is the Limit – but Where does it End?" *Proceedings of the Forty-Eighth Colloquium on the Law of Outer Space*. IISL, Fukuoka, 2005. p. 92.

[20] Indeed, as observed by Armand D. Roth, "la question est en réalité (…) plus politique que technique et la prolifération des théories en la matière semble inversement proportionnelle aux perspectives de solution". *La Prohibition de l'Appropriation et les Régimes d'Accès aux Espaces Extra-Terrestres*. Paris, France: Presses Universitaires de France, 1992. p. 99.

territorial State, which, otherwise, could be entitled to promote protective alternatives to safeguard its national interests in extreme circumstances.

Accordingly, it is hereby proposed that the passage of space objects through foreign air space should not be considered peaceful if conducted in violation of International Law, disrespecting the sovereign rights of the territorial State or implicating unjustified dangers to its national security or the wellbeing of the local population and environment.

Considering that an international problem is under examination, affecting each and every State, the solution should preferably be achieved through a new international convention, at least by amendment or protocol to the Space Treaty or the Chicago Convention, in order to guarantee proper legal security. Other regulatory options, including a United Nations General Assembly Resolution, while internationally relevant, could not prevent questions related to the consolidation of a consuetudinary rule that authorizes passage rights for space activities.

Through careful treaty regulation, this new legal regime may be clearly drafted, assuring that the exploration of outer space continues to be conducted for the benefit of all mankind. Likewise, by eliminating an important lacuna in Air Law and Space Law, the potential for international disputes would be minimized, safeguarding peace in international relations, undeniably the major objective of International Law.

One may assert that only through a compromise, which reaches for the common place between conflicting approaches, will it be possible to clarify the international rules applicable to air space and outer space.

It is strongly believed that the delimitation of the frontier between air space and outer space should be regarded not as a scientific but as a legal problem, with practical implications, due to inexorable progress of science. Therefore, a multilateral legal solution should be pursued. Through debates, proposals may be discussed and alternatives may be endeavored. It is about time to revisit this important topic that, after so many years since the beginning of the Space Age, still awaits a final, clear and proper answer by the international community.

Bibliography

Books and Legislation

Abad, Rodolfo R. *El Estado en el Desarrollo de la Comunidad*. Buenos Aires, Argentina: Club de Lectores, 1968.

Abeyratne, Ruwantissa I. R. *Frontiers of Aerospace Law*. Burlington, USA: Ashgate, 2002.

Abranches, C. A. Dunshee de. *Espaço Exterior e Responsabilidade Internacional*. Rio de Janeiro, Brazil: Livraria Freitas Bastos, 1964.

Accioly, Hildebrando; Nascimento E Silva, G. E. do; and Casella, Paulo Borba. *Manual de Direito Internacional Público*. 16. ed. São Paulo, Brazil: Saraiva, 2008.

Accioly, Hildebrando. *Tratado de Direito Internacional Público*. 3. ed. São Paulo, Brazil: Quartier Latin, 2009.

Albuquerque, Newton de Menezes. *Teoria Política da Soberania*. Belo Horizonte, Brazil: Mandamentos, 2001.

Alburquerque Mello, Celso D. de. *Curso de Direito Internacional Público*. 14. ed. São Paulo, Brazil: Editora Renovar, 2002.

Amaral Junior, Alberto do. *Introdução ao Direito Internacional Público*. São Paulo, Brazil: Atlas, 2008.

Andem, Maurice N. *International Legal Problems in the Peaceful Exploration and Use of Outer Space*. Rovaniemi, Finland: University of Lapland, 1992.

Anderson, John D., Jr. *A History of Aerodynamics*. Cambridge, England: Cambridge, 1997.

Araújo, Luis Ivani de Amorim. *Curso de Direito Internacional Público*. 6. ed. Rio de Janeiro, Brazil: Forense, 1988.

Aréchaga, Eduardo Jiménez (dir.). *Derecho Internacional Publico*. Montevidéu, Uruguay: Fundación de Cultura Universitaria, 1990.

Aust, Anthony. *Handbook of International Law*. Cambridge, England: Cambridge University Press, 2005.

Azambuja, Darcy. *Teoria Geral do Estado*. 39. ed. Rio de Janeiro, Brazil: Globo, 1998.

Azevedo, Oldemar. *Soberania: Noções e Fundamentos*. São Paulo, Brazil: Lex, 1970.

Badie, Bertrand. *Um Mundo sem Soberania: os Estados entre o Artifício e a Responsabilidade*. Lisbon, Portugal: Insituto Piaget, 1999.

Badie, Bertrand e Vidal, Dominique (dir.). *El Estado del Mundo: Anuario Económico Geopolítico Mundial*. Madrid, Spain: Akal, 2009.

Bahia, Luiz Alberto. *Soberania, Guerra e Paz*. Rio de Janeiro, Brazil: Zahar, 1978.

Barberis, Julio A. *Formación Del Derecho Internacional*. Buenos Aires, Argentina: Ábaco, 1994.

© Springer International Publishing Switzerland 2015

O.O. Bittencourt Neto, *Defining the Limits of Outer Space for Regulatory Purposes*, SpringerBriefs in Space Development,
DOI 10.1007/978-3-319-16685-8

Bate, Roger R.; Mueller, Donald D. and White, Jerry E. *Fundamentals of Astrodynamics*. New York, USA: Dover, 1971.
Bauza, Araújo. *Hacia un Derecho Astronáutico*. Montevidéu, Uruguay: Roque Desalma, 1957.
Belfort De Mattos, José Dalmo Fairbanks. *Direito Internacional*. São Paulo, Brazil: Saraiva, 1979.
Benkö, Marietta; De Graaff, Willem and Reijnen, Gijsbertha C. M. *Space Law and the United Nations*. Dordrecht, the Netherlands: Martinus Nijhoff, 1985.
Benkö, Marietta and Schrogl, Kai-Uwe (eds.). *International Space Law in the Making: Current Issues in the UN Committee on the Peaceful Uses of Outer Space*. Paris, France: Frontières, 1993.
Benkö, Marietta and Plescher, Engelbert. *Space Law – Reconsidering the Definition/Delimitation Question and the Passage of Spacecraft Through Foreign Airspace*. The Hague, The Netherlands: Eleven, 2013.
Berliner, Don. *Aviation: Reaching for the Sky*. Minneapolis, USA: The Oliver Press, 1997.
Beviláqua, Clóvis. *Direito Público Internacional*. Tomo I. Rio de Janeiro, Brazil: Livraria Francisco Alves, 1911.
Bhatt, S. *International Aviation and Outer Space Law and Relations: Reflections on Future Trends*. New Deli, India: Asian Institute of Transport Development, 1996.
Bhayana, Harnam. *International Law in the Regime of Outer Space*. Calcutta, India: R. Cambray & Co. Pvt., 2001.
Bittencourt Neto, Olavo de O. *Responsabilidade Internacional dos Estados no Direito Espacial: Brasil como Estado Lançador*. São Paulo, Brazil: Faculdade de Direito da Universidade de São Paulo, Brazil, 2008.
Bittencourt Neto, Olavo de O. *Direito Espacial Contemporâneo: Responsabilidade Internacional*. Curitiba: Juruá, 2011.
Blainey, Geoffrey. *Uma Breve História do Século XX*. São Paulo, Brazil: Fundamento Educacional, 2008.
Bobbio, Norberto; Matteucci, Nicola; and Pasquino, Gianfranco. *Dicionário de Política*. 13. ed. Brasília: UNB, 2007.
Bodin, Jean. *On Sovereignty*. Book 8, Chapter 1. Cambridge, England: 2007.
Brown, E. D. *The International Law of the Sea*. Vol. I. Aldershot, USA: Dartmouth, 1994.
Brownlie, Ian. *Principles of Public International Law*. 6. ed. Oxford, England: Oxford University Press, 2003.
Brunner, Christian; Ulrich, Silvia e Lauer, Andrea. *Funktionen des Rechts in der Pluralistischen Wissensgesellschaft*. Vienna, Austria: Böhlau, 2007.
Brzezinski, Matthew *La Conquista del Espacio*. Buenos Aires, Argentina: El Ateneo, 2008.
Burdeau, Georges. *O Estado*. São Paulo, Brazil: Martins Fontes, 2005.
Burrows, William E. *This New Ocean – The Story of the First Space Age*. New York, USA: Modern Library, 1999.
Campbell, Joseph. *O Poder do Mito*. São Paulo, Brazil: Palas Athena, 1990.
Casella, Paulo Borba. *Direito Internacional, Terrorismo e Aviação Civil*. São Paulo, Brazil: Quartier Latin, 2006.
Casella, Paulo Borba. *Fundamentos do Direito Internacional Pós-Moderno*. São Paulo, Brazil: Quartier Latin, 2008.
Casella, Paulo Borba. *Direito Internacional dos Espaços*. São Paulo, Brazil: Atlas, 2009.
Cassese, Antonio. *International Law*. 2. ed. Oxford, England: Oxford University Press, 2005.
Chanute, Octave. *Progress in Flying Machines*. Mineola, USA: Dover, 1997.
Chaumont, Charles. *Le Droit de l'Espace*. Paris, France: Presses Universitaires de France, 1960.
Cheng, Bing. *Studies on International Space Law*. Oxford, England: Clarendon Pr, 1998.
Cheng, Chia-Jui (ed.). *The Use of Air Space and Outer Space for All Mankind in the 21st Century*. The Hague, the Netherlands: Kluwer, 1995.
Cheuiche, Alcy. *Santos-Dumont*. Porto Alegre, Brazil: L&PM, 2009.
Christol, Carl Q. *Space Law: Past, Present and Future*. The Hague, the Netherlands: Kluwer Academic Publishers, 1991.
Clarke, Arthur C. *A Exploração do Espaço*. São Paulo, Brazil: Melhoramentos, 1959.
Costadoat, Pasini. *El Espacio Aéreo*. Buenos Aires, Argentina: Depalma, 1955.

Crawford, James. *The Creation of States in International Law*. 2. ed. New York, USA: Oxford University Press, 2006.
Crouch, Tom D. *Asas*. Rio de Janeiro, Brazil: Record, 2008.
Csabafi, Imre Anthony. *The Concept of State Jurisdiction in International Space Law*. The Hague, the Netherlands: Martinus Nijhoff, 1971.
Dallari, Dalmo de Abreu. *Elementos da Teoria Geral do Estado*. 19. ed. São Paulo, Brazil: Saraiva, 1995.
Dallari, Pedro Bohomoletz de Abreu Dallari (coord.). *Relações Internacionais: Múltiplas Dimensões*. São Paulo, Brazil: Aduaneiras, 2004.
De Almeida, Francisco Ferreira. *Direito Internacional Público*. 2. ed. Coimbra, Portugal: Coimbra Editora, 2003.
Del Vecchio, Giorgio. *Teoria do Estado*. São Paulo, Brazil: Saraiva, 1997.
Delorme, Jean. *La Mer et L'Espace*. Paris, France: Hachette, 1966.
Dickson, Paul. *Sputnik: the Shock of the Century*. Waterville, USA: G. K. Hall & Co., 2001.
Diederiks-Verschoor, I. H. Ph. *An Introduction to Air Law*. 5. ed. Dordrecht, the Netherlands: Kluwer Academic Publishers, 1993.
Diederiks-Verschoor, I. H. Ph. *An Introduction to Space Law*. 2. ed. Dordrecht, the Netherlands: Kluwer Academic Publishers, 1999.
Direito Espacial – Coletânea De Covenções, Atos Internacionais E Diversas Disposições Legais Em Vigor. Agência Espacial Brasileira e Sociedade Brasileira de Direito Aeroespacial. Rio de Janeiro, Brazil: Imprensa Nacional, 2003.
Dolinger, Jacob. *Direito Internacional Privado – Parte Geral*. 5. ed. Rio de Janeiro, Brazil: Renovar, 2000.
Duguit, Léon. *Soberania y Libertad*. Buenos Aires, Argentina: Tor, 1943.
Duguit, Léon. *Fundamentos do Direito*. São Paulo, Brazil: Martin Claret, 2009.
Dupuy, Pierre-Marie. *Droit International Public*. 8. ed. Paris, France: Dalloz, 2006.
Dupuy, René-Jean. *O Direito Internacional*. Coimbra, Portugal: Livraria Almedina, 1993.
Duso, Giuseppe. *O Poder: História da Filosofia Política Moderna*. Petrópolis: Vozes, 2005.
Dussel, Enrique. *20 Teses de Política*. São Paulo, Brazil: Expressão Popular, 2007.
Einstein, Albert. *Relativity: the Special and the General Theory*. New York, USA: Dover, 2010.
Epiphanio, José Carlos Neves; Novo, Evlyn Márcia Leão de Moraes and Machado, Luiz Augusto Toledo. *Espaço*. São Paulo, Brazil: Blucher, 2010.
Faro Junior, Luiz P. F. de. *Direito Internacional Público*. 3. ed. Rio de Janeiro, Brazil: Haddad, 1960.
Fauchille, Paul. *Le Domaine Aéreian et Le Régime Juridique de les Aerostats*. Paris, France: Pedone, 1901.
Fawcett, J. E. S. *International Law and the Uses of Outer Space*. Manchester, England: University Press, 1969.
Fawcett, J. E. S. *Outer Space: New Challenges to Law and Policy*. Oxford, England: Clarendon Press, 1984.
Ferrajoli, Luigi. *A Soberania no Mundo Moderno*. São Paulo, Brazil: Martins Fontes, 2007.
Ferrer, Manuel Augusto. *Derecho Espacial*. Buenos Aires, Argentina: Plus Ultra, 1976.
Fischer, David I. *Prior Consent to International Direct Satellite Broadcasting*. Dordrecht, the Netherlands: Martinus Nijhoff, 1990. p. 159.
Fleiner-Geister, Thomas. *Teoria Geral do Estado*. São Paulo, Brazil: Martins Fontes, 2006.
Foucher, Michel. *Obsessão por Fronteiras*. São Paulo, Brazil: Radical Livros, 2009.
Freitas, Paulo Henrique de Souza. *Responsabilidade Civil no Direito Aeronáutico*. São Paulo, Brazil: Juarez de Oliveira, 2003.
Gaddis, John Lewis. *História da Guerra Fria*. Rio de Janeiro, Brazil: Nova Fronteira, 2006.
Gaggero, Eduardo D. *Uruguay ante el Derecho Espacial*. Montevidéu, Uruguay: Don Orione, 1975.
Gainor, Chris. *To a Distant Day: the Rocket Pioneers*. New York, USA: Random House, 2008.
Gál, Gyúla. *Space Law*. Leiden, the Netherlands: A. W. Sijthoff, 1969.
Gangale, Thomas. *The Development of Outer Space: Sovereignty and Property Rights in International Space Law*. Santa Barbara, USA: Praeger, 2009.
Gemael, Camil and De Andrade, José Bittencourt. *Geodésia Celeste*. Curitiba: UFPR, 2004.

Goedhart, Robert F. A. *The Never Ending Dispute: Delimitation of Air Space and Outer Space*. Paris, France: Frontières, 1996.
Goes Filho, Synesio Sampaio. *Navegantes, Bandeirantes e Diplomatas: um Ensaio sobre a Formação das Fronteiras do Brasil*. São Paulo, Brazil: Martins Fontes, 2001.
Goh, Gérardine Meishan. *Dispute Settlement in International Space Law: a Multi-Door Courthouse for Outer Space*. Dordrecht, the Netherlands: Martinus Nijhoff, 2007.
Goldman, Nathan C. *Space Policy: an Introduction*. Iowa, USA: Iowa State, 1992.
Golovine, M. N. *Conflict in Space: a Pattern of War in a New Dimension*. London, England: Temple Press, 1962.
Górbiel, Andrzej e Palyga, Edward J. *International Space Law Miscellanea: Liber Amicorum Honouring Professor Dr. Andrzej Górbiel in his 65th Anniversary*. Warsaw, Poland: Andrzej Frycz Modrzewski Foundation, 1995.
Górbiel, Andrzej. *Legal Definition of Outer Space*. Lodz, Poland: Uniwersytet Lódzki, 1980.
Grotius, Hugo. *O Direito da Guerra e da Paz (De Jure Belli ac Pacis)*. 2. ed. Ijuí, Brazil: Unijuí, 2004.
Grotius, Hugo. *Mare Liberum*. Leiden, the Netherlands: Brill, 2009.
Guerra, Sidney and Silva, Roberto Luiz (orgs.). *Soberania: Antigos e Novos Paradigmas*. Rio de Janeiro, Brazil: Freitas Bastos, 2004.
Guerra, Sidney. *Direito Internacional Público*. 2. ed. Rio de Janeiro, Brazil: Freitas Bastos, 2005.
Haanappel, Peter P. C. *The Autonomy of Air and Space Law, Visited and Revisited*. Leiden, the Netherlands: Universiteit Leiden, 1997.
Haanappel, Peter P. C. *The Law and Policy of Air Space and Outer Space: a Comparative Approach*. Dordrecht, the Netherlands: Kluwer Academic Publishers, 2003.
Haley, Andrew G. *Space Law and Government*. New York, USA: Appleton-Century-Crofts, 1963.
Hallion, Richard P. *Taking Flight: Inventing the Aerial Age from Antiquity through the First World War*. New York, USA: Oxford, 2003.
Hardesty, Von and Eisman, Gene. *Epic Rivalry – The Inside Story of the Soviet and American Space Race*. Washington, D.C., USA: National Geographic, 2007.
Hinsley, F. H. *El Concepto de Soberania*. Barcelona, Spain: Labor, 1975.
Hobe, Stephan, Schmidt-Tedd, Bernhard and Schrogl, Kai-Uwe (eds.). *Cologne Commentary on Space Law*. Volume 1 – Outer Space Treaty. Cologne, Germany: Carl Heymanns, 2009.
Howard, Fred. *Wilbur and Orville: a Biography of the Wright Brothers*. Mineola, USA: Dover, 1998.
Hurwitz, Bruce A. *The Legality of Space Militarization*. Amsterdam, the Netherlands: North-Holland, 1986.
Hurwitz, Bruce A. *State Liability for Outer Space Activities: in Accordance with the 1972 Convention on International Liability for Damage Caused by Space Objects*. Dordrecht, the Netherlands: Kluwer Academic Publishers, 1992.
Huzek, Carlos Roberto. *Curso de Direito Internacional Público*. 4. ed. São Paulo, Brazil: LTr, 2003.
Hyman, William A. *Magna Carta of Space*. Amherst, USA: Amherst Press, 1962.
Impallomeni, Elisabeth Back. *Spazio Cósmico e Corpi Celesti nell'Ordinamento Internazionale*. Padova, Italy: CEDAM, 1983.
Jackson, Robert and Sørensen, Georg. *Introdução às Relações Internacionais*. Rio de Janeiro, Brazil: Jorge Zahar, 2007.
Jagota, S. P. *Maritime Boundaries*. Dordrecht, the Netherlands: Martinus Nijhoff, 1985.
Janis, Mark W. *An Introduction to International Law*. 2. ed. USA: Aspen Law & Business, 1993.
Janis, Mark W. and Noyes, John E. *International Law – Cases and Commentary*. St. Paul, USA: West Group, 2001.
Jasani, Bhupendra (ed.). *Outer Space – a New Dimension of the Arms Race*. London, England: Taylor & Francis, 1982.
Jasani, Bhupendra (ed.). *Peaceful and Non-Peaceful Uses of Space*. New York, USA: Taylor & Francis, 1991.
Jasentuliyana, Nandasiri and Lee, Roy S. K. (eds.). *Manual on Space Law*. Volume I. New York, USA: Oceana, 1979.
Jasentuliyana, Nandasiri (coord.). *Space Law: Development and Scope*. New York, USA: Praeger Publishers, 1992.

Jasentuliyana, Nandasiri. *Perspectives in International Law.* The Hague, the Netherlands: Kluwer, 1995.
Jasentuliyana, Nandasiri. *International Space Law and the United Nations*. The Hague, the Netherlands: Kluwer, 1999.
Jellinek, Georg. *Teoria General Del Estado*. Mexico City, Mexico: FCE, 2000.
Jenks, C. Wilfred. *Space Law*. New York, USA: Frederick A. Praeger, 1965.
Johnson, David. *Rights in Air Space.* Manchester, England: Manchester University Press, 1965.
Johnson-Freese, Joan. *Space as a Strategic Asset*. New York, USA: Columbia, 2007.
Kant, Immanuel. *The Critique of Practical Reason*. Radford, USA: A & D, 2008.
Kaplan, Morton A. and Katzenbach, Nicholas de B. *Fundamentos Políticos do Direito Internacional.* Rio de Janeiro, Brazil: Zahar, 1964.
Kayser, Valérie. *Launching Space Objects: Issues of Liability and Future Prospects.* Dordrecht, the Netherlands: Kluwer Academic Publishers, 2001.
Kelsen, Hans. *Principles of International Law.* 2. ed. New York, USA: Holt, Rinehart and Winston, 1967.
Kelsen, Hans. *Teoria Geral do Estado e do Direito*. São Paulo, Brazil: Martins Fontes, 2005.
Kemp, Kenny. *Destination Space: How Space Tourism is Making Science Fiction a Reality*. London, England: Virgin Books, 2007.
Kerrest, Armel (org.). *L'Adaptation du Droit de l'Espace a ses Nouveaux Defis: Mélanges en l'Honneur de Simone Courteix*. Paris, France: Pedone, 2007.
Kisch, John. *The Law of International Spaces*. Leiden, the Netherlands: Sijthoff, 1973.
Kroell, Joseph. *Traité de Droit International Public Aérien.* Tome I. Paris, France: Les Éditions Internationales, 1934.
Lacerda, J. C. Sampaio de. *Curso de Direito Comercial Marítimo e Aeronáutico (Direito Privado da Navegação*). Rio de Janeiro, Brazil: Freitas Bastos, 1957.
Lachs, Manfred. *The Law of Outer Space*. Leiden, the Netherlands: Martinus Nijhoff, 2010.
Lafferranderie, Gabriel (ed.). *Outlook on the Space Law over the Next 30 Years: Essays Published for the 30th Anniversary of the Outer Space Treaty*. The Hague, the Netherlands: Kluwer, 1997.
Ley, Willy. *Rockets, Missiles and Space Travel*. 2. ed. New York, USA: Viking, 1961.
Levi, Werner. *Contemporary International Law*. 2. ed. Boulder, USA: Westview Press, 1991.
Litrento, Oliveiros. *A Ordem Internacional Contemporânea: um Estudo da Soberania em Mudança.* Porto Alegre, Brazil: Sergio Antonio Fabris, 1991.
Litrento, Oliveiros. *Curso de Direito Internacional Público*. 4. ed. Rio de Janeiro, Brazil: Forense, 2001.
Long, Colin D. *Telecommunications Law and Practice.* 2. ed. London, England: Sweet & Maxwell, 1995.
Lopes, Mario Justo. *La Sobernia*. Buenos Aires, Argentina: Cooperadora de Derecho y Ciencias Sociales, 1967.
Lyall, Francys e Laursen, Paul B. *Space Law: a Treatise*. Farnham, England: Ashgate, 2009.
Lysén, Göran. *State Responsibility and International Liability of States for Lawful Acts: a Discussion of Principles*. Gottemburg, Sweden: Iustus Förlag, 1997.
Macedo, Paulo Emílio Borges de. *O Nascimento do Direito Internacional*. São Leopoldo, Brazil: Unisinos, 2009.
Malanczuk, Peter. *Akehurst's Modern Introduction to International Law*. 7. ed. London, England: Routledge, 1997.
Malberg, R. Carré de. *Teoria General del Estado*. Mexico City, Mexico: FCE, 2001.
Malin, Stuart. *The Greenwich Guide to the Planets*. London, England: George Philip, 1989.
Masson-Zwaan, Tanja (ed.). *Space Law: Views of the Future.* Deventer, the Netherlands: Kluwer, 1988.
Masson-Zwaan, Tanja and De Leon, Pablo Mendes (eds.). *Air and Space Law: De Lege Ferenda – Essays in Honour of Henri A. Wessenbergh*. Dordrecht, the Netherlands: Martinus Nijhoff, 1992.
Matias, Eduardo Felipe P. *A Humanidade e Suas Fronteiras: do Estado Soberano à Sociedade Global*. São Paulo, Brazil: Paz e Terra, 2005.
Matte, Nicolas Mateesco. *Deux Frontières Invisibles: De la Mer Territoriale à l'Air 'Territorial'*. Paris, France: Pedone, 1965.
Matte, Nicolas Mateesco. *Aerospace Law: from Scientific Exploration to Commercial Utilization*. Toronto, Canada: Carswell, 1977.
Mattos, Adherbal Meira. *Direito Internacional Público*. 2. ed. Rio de Janeiro, Brazil: Renovar, 2002.
Mattos, Adherbal Meira. *O Novo Direito do Mar*. 2. ed. Rio de Janeiro, Brazil: Renovar, 2008.

Mcdougal, Myres S., Lasswell, Harold D. and Vlasic, Ivan I. *Law and Public Order in Space*. New Haven, USA: Yale University Press, 1963.
Mcdougall, Walter A. ... *The Heavens and the Earth – A Political History of the Space Age*. New York, USA: Basic Books, 1985.
Mcnair, Arnold Duncan. *The Law of the Air*. 3. ed. London, England: 1964.
Menezes, Aderson de. *Teoria Geral do Estado*. 5. ed. Rio de Janeiro, Brazil: Forense, 1993.
Mercadante, Araminta and Magalhães, José Carlos de (orgs.) *Reflexões sobre os 60 Anos da ONU*. Ijuí, Brazil: Unijuí, 2005.
Milde, Michael. *International Air Law and ICAO*. Utrecht, the Netherlands: Eleven International Publishing, 2008.
Miranda, Jorge. *Curso de Direito Internacional Público*. 3. ed. Parede, Portugal: Princípia, 2008.
Monateri, Pier Giuseppe (ed.). *Methods of Comparative Law*. Cheltenham, UK: Edward Elgar Publishing, 2012.
Monserrat Filho, José. *Direito e Política na Era Espacial*. Rio de Janeiro, Brazil: Vieira & Lent, 2007.
Montesquieu, Charles de. *Do Espírito das Leis*. São Paulo, Brazil: Martin Claret, 2002.
Morais, Antônio Manuel Alves. *Gravitação & Cosmologia: uma Introdução*. São Paulo, Brazil: Livraria da Física, 2010.
Morenoff, Jerome. *World Peace Through Space Law*. Charlottesville, USA: Michie Co., 1967.
Morris, Clarence (org.). *Os Grandes Filósofos do Direito*. São Paulo, Brazil: Martins Fontes, 2002.
Mourão, Ronaldo Rogério de Freitas. *Astronáutica – do Sonho à Realidade: História da Conquista Espacial*. Rio de Janeiro, Brazil: Bertrand Brasil, 1999.
Murchison, John Taylor. *The Contiguous Air Space Zone in International Law*. Ottawa, Canada: Department of National Defence, 1957.
Newton, Isaac. *Principia – Princípios Matemáticos de Filosofia Natural (Livro III)*. São Paulo, Brazil: Folha de São Paulo, Brazil, 2010.
Nguyen, Quoc Dinh; Pellet, Alain; Daillier, Patrick. *Direito Internacional Público*. Lisbon, Portugal: Fundação Calouste Gulbenkian, 1999.
Nogueira, Salvador. *Rumo ao Infinito – Passado e Futuro da Aventura Humana na Conquista do Espaço*. São Paulo, Brazil: Globo, 2005.
Nogueira, Salvador. *Conexão Wright – Santos-Dumont: a Verdadeira História da Invenção do Avião*. Rio de Janeiro, Brazil: Record, 2006.
Nordquist, Myron H., Koh, Tommy T. B. and Moore, John Norton (ed.). *Freedom of Seas, Passage Rights and the 1982 Law of the Sea Convention*. Leiden, the Netherlands: Martinus Nijhoff, 2009.
Novaes, Adauto (org.). *A Crise do Estado-Nação*. Rio de Janeiro, Brazil: Civilização Brasileira, 2003.
Ogunbanwo, Ogunsola O. *International Law and Outer Space Activities*. The Hague, the Netherlands: Martinus Nijhoff, 1975.
Ovidius. *Metamorfoses*. Vol. II. Lisbon, Portugal: Vegas, 2008.
Pacheco, José da Silva. *Comentários ao Código Brasileiro de Aeronáutica*. 4. ed., Rio de Janeiro, Brazil: Forense, 2006.
Painter, David S. *The Cold War: an International History*. New York, USA: Routledge, 1999.
Pancracio, Jean-Paul. *Droit International des Espaces*. Paris, France: Armand Colin, 1997.
Paubel, Emerson Faria Cabral. *Propulsão e Controle de Veículos Aeroespaciais*. Florianópolis, Brazil: UFSC, 2002.
Paupério, A. Machado. *Teoria Democrática da Soberania*. Rio de Janeiro, Brazil: Forense, 1997.
Pellegrino, Carlos Roberto. *História da Ordem Internacional*. São Paulo, Brazil: Brasiliense, 1988.
Pessoa, Epitácio. *Projeto de Código de Direito Internacional Público*. Rio de Janeiro, Brazil: Imprensa Nacional, 1911.
Pisier, Evelyne. *História das Ideias Políticas*. Barueri, Brazil: Manole, 2004.
Prescott, Victor and Triggs, Gillian T. *International Frontiers and Boundaries: Law, Politics and Geography*. Dordrecht, the Netherlands: Martinus Nijhoff, 2008.
Reijnen, Gijsbertha Cornelia Maria. *Legal Aspects of Outer Space*. Utrecht, the Netherlands: Drukkerij Elinkwijk, 1977.
Reinhardt, Dean. *The Vertical Limit of State Sovereignty*. Institute of Air and Space Law, McGill University. Montréal, Canada: 2005.

Reuter, Paul. *Droit International Public*. 4. ed. Paris, France: Presses Universitaires de France. 1973.
Reuter, Paul. *Instituições Internacionais*. Lisbon, Portugal: Edições Rolim, 1975.
Reuter, Paul. *Introducción al Derecho de los Tratados*. 2. ed. Mexico City, Mexico: FCE, 2001.
Reynolds, Glenn H and Merges, Robert P. *Outer Space: Problems of Law and Policy*. 2. ed. Boulder, USA: Western Press, 1998.
Rezek, Francisco. *Direito Internacional Público*. 10. ed. São Paulo, Brazil: Saraiva, 2007.
Rochère, Jacqueline Dutheil de la (ed.). *Droit de l'Espace*. Paris, France: Pedone, 1988.
Roth, Armand D. *La Prohibition de l'Apropriation et les Regimes d'Accès aux Espaces Extra-Terrestres*. Paris, France: Presses Universitaires de France, 1992.
Rothwell, Donald R. and Stephens, Tim. *The International Law of the Sea*. Oxford, England: Hart Publishing, 2010.
Rousseau, Charles. *Droit International Public*. 11. ed. Paris, France: Dalloz, 1987.
Sadeh, Eligar (org.). *Space Politics and Policy – An Evolutionary Perspective*. Dordrecht, the Netherlands: Kluwer Academic Publishers, 2002.
Saha, Kshudiram. *The Earth's Atmosphere: its Physics and Dynamics*. New York, USA: Springer, 2008.
Salcedo, Juan Antonio Carrillo. *Soberanía del Estado y Derecho Internacional*. 2. ed. Madrid, Spain: Editorial Tecnos, 1976.
Salinas, Luis Tapia. *Manual de Derecho Aeronáutico*. Barcelona, Spain: Bosch Casa Editorial, 1944.
Santos, Milton. *A Natureza do Espaço: Técnica e Tempo, Razão e Emoção*. 4. ed. São Paulo, Brazil: Edusp, 2009.
Santos-Dumont, Alberto. *O Que Eu Vi, o Que Nós Veremos*. São Paulo, Brazil: Hedra, 2000.
Schlesinger, Rudolf B.; Baade, Hans W.; Herzog, Peter E. and Wise, Edward W. *Comparative Law*. 6. ed. New York, USA: 1998.
Schofield, Martin B. *Control of Outer Space*. New York, USA: 1959.
Shaw, Malcolm N. *International Law*. 3. ed. Cambridge, England: Cambridge University Press, 2003.
Singh, J. N. *Outer Space, Outer Sea, Outer Land and International Law*. Nova Deli, India: Harnam Publications, 1987.
Slomanson, William R. *Fundamental Perspectives on International Law*. 4. ed. Belmont, USA: Wadsworth/Thomson Learning, 2003.
Smith, Milton L. *International Regulation of Satellite Communication*. Dordrecht, the Netherlands: Martinus Nijhoff, 1990.
Soares, Albino de Azevedo. *Lições de Direito Internacional Público*. 4. ed. Coimbra, Portugal: Coimbra Editora Limitada, 1988.
Soares, Guido Fernando Silva. *Curso de Direito Internacional Público*. São Paulo, Brazil: Atlas, 2002.
Speciali, Raymond C. *Fundamentals of Aviation Law*. New York, USA: McGraw-Hill, 2006.
Stadtmuller, Georg. *História del Derecho Internacional Publico*. Part I. Madrid, Spain: Aguilar, 1961.
Stephanides, Menelaos. *Prometeu, os Homens e Outros Mitos*. São Paulo, Brazil: Odysseus, 2008.
Taubman, Phil. *Secret Empire: Eisenhower, the CIA, and the Hidden Story of American Space Espionage*. New York, USA: Simon & Schuster, 2003.
Tunkin, G. (coord.). *Curso de Derecho Internacional: Manual*. Moscow, Russia: Progresso, 1979.
Uprety, Kishor. *The Transit Regime for Landlocked States*. Washington D.C., USA: the World Bank, 2006.
Van Kleffens, Eeco. *Sovereignty in International Law: Five Lectures*. Recueil de Cours, 082. The Hague, the Netherlands: Martinus Nijhoff, 1953.
Vattell, Emer de. *O Direito das Gentes*. Brasilia, Brazil: IPRI, 2004.
Vázquez, Modesto Seara. *Derecho y Política en el Espacio Cósmico*. Mexico City, Mexico: Universidad Nacional Autónoma de México, 1986.
Verdross, Alfred. *Volkerrecht*. 5. ed. Vienna, Austria: Springer Verlag, 1964.
Videla Escalada, Federico N. *Aeronautical Law*. Alpheen aan den Rijn, the Netherlands: Sijthoff & Noordhoff, 1979.
Virally, Michel. *El Devenir del Derecho Internacional*. Mexico City, Mexico: FCE, 1998.
Vitória, Francisco de. *Os Índios e o Direito da Guerra*. Ijuí, Brazil: Unijuí, 2006.
Wadegaonkar, Damodar. *The Orbit of Space Law*. London, England: Stevens & Sons, 1984.

Wagner, Wenceslas J. *International Air Transportation as Affected by State Sovereignty.* Brussels, Belgium: Bruylant, 1970.
Wassenbergh, Henri. *Principles of Outer Space in Hindsight.* Dordrecht, the Netherlands: Kluwer Academic Publishers, 1991.
Winter, Frank H. *Rockets into Space.* Cambridge, USA: Harvard University Press, 1990.
Winter, Othon Cabo and Prado, Antonio Fernando Bertachini de Almeida. *A Conquista do Espaço: do Sputnik à Missão Centenário.* São Paulo, Brazil: Editora Livraria da Física, 2007.
Wright, Orville. *How We Invented the Airplane: an Illustrated History.* Mineola, USA: Dover, 1988.

Papers

Aaronson, Michael. “Space Law”. *International Relations*, vol. I, no. 9. New York, USA, April/1959.
Becker, Loftus. “Major Aspects of the Problem of Outer Space”. *Bulletin of the Department of State.* Washington, USA, 1958.
Bevilacqua, Luiz de Gonzaga. “A Contribution to the Problem of Space Law Establishing a Technical and Practical Limit to Political Sovereignty in Space”. *Proceedings of the First Colloquium on the Law of Outer Space.* IISL, The Hague, 1957.
Cheng, Chia-Jui. “New Sources of International Air Law”, *In*: Cheng, Chia-Jui (ed.). *The Use of Air Space and Outer Space for All Mankind in the 21st Century.* The Hague, the Netherlands: Kluwer, 1995.
Cooper, John Cobb. “Roman Law and the Maxim Cujus Est Solum in International Air Law”. *McGill L. J.*, n. 23. Montréal, Canada: 1955.
Dallari, Pedro Bohomoletz de Abreu. “Aspectos Jurídicos da Formação e da Gestão do Território Nacional: o Caso Brasileiro”, *In*: Pedro Bohomoletz de Abreu Dallari (coord.). *Relações Internacionais: Múltiplas Dimensões.* São Paulo, Brazil: Aduaneiras, 2004.
Danilenko, Gennady M. “The Boundary Between Air Space and Outer Space in Modern International Law: Delimitation on the Basis of Customary Law”. *Proceedings of the Twenty-Sixth Colloquium on the Law of Outer Space.* IISL, Budapest, 1983.
Dembling, Paul G. “Principles Governing the Activities of States in the Exploration and Use of Outer Space, Including the Moon and Other Celestial Bodies”, *In:* Jasentuliyana, Nandasiri and Lee, Roy S. K. (eds.). *Manual on Space Law.* Volume I. New York, USA: Oceana, 1979.
De Vries, Walter W. C. “Boundaries in Space? Some Theories and Problems Regarding the Definition, Delimitation and/or Demarcation of Outer Space”. *Thesauros Acroasium.* Institution of International Public Law and International Relations of Thessaloniki, Vol. XIV. Thessaloniki, Greece, 1985.
De Vries, Walter W. C. “The Creation of a Concepto f the Law of Outer Space”, *In:* Masson-Zwaan, Tanja (ed.). *Space Law: Views of the Future.* Deventer, the Netherlands: Kluwer, 1988.
Diederiks-Verschoor, I. H. “Differences Between Air Law and Space Law”. *Recueil des Cours*, 172. The Hague, the Netherlands, 1981.
Djalal, Hasjim. “Remarks on the Concept of ‘Freedom of the Seas”, *In:* Nordquist, Myron H., Koh, Tommy T. B. and Moore, John Norton (ed.). *Freedom of Seas, Passage Rights and the 1982 Law of the Sea Convention.* Leiden, the Netherlands: Martinus Nijhoff, 2009.
Doyle, S. E. “Concepts of Space Law before Sputnik”. *Proceedings of the Fortieth Colloquium on the Law of Outer Space.* IISL, Turin, 1997.
Fernández-Brital, Oscar. “UN-COPUOS Legal Subcommittee March-April 1993 Meeting and the Limit between Air Space and Outer Space”. *Proceedings of the Twenty-Seventh Colloquium on the Law of Outer Space.* IISL, Jerusalem, 1983.
Freeland, Steven and Jakhu, Ram. “Article II”, *In:* Hobe, Stephan, Schmidt-Tedd, Bernhard, Schrogl, Kai-Uwe (eds.). *Cologne Commentary on Space Law.* Volume 1 – Outer Space Treaty. Cologne, Germany: Carl Heymanns, 2009.

Gál, Gyúla. "The Question of Delimitation – After Twenty Years". *Proceedings of the Twenty-Second Colloquium on the Law of Outer Space*. IISL, Munich, 1979.

Gál, Gyúla. "Fundamental Links and Conflicts between Legal Rules of Air and Space Flights". *Proceedings of the Twenty-Sixth Colloquium on the Law of Outer Space*. IISL, Budapest, 1983.

Gál, Gyúla. "Thirty Years of Functionalism". *Proceedings of the Fortieth Colloquium on the Law of Outer Space*. IISL, Turin, 2005.

Galloway, Eilene. "Legal Aspects of International Cooperation in Space: Area and Functional Concepts in Defining Outer Space". *Proceedings of the Twenty-Sixth Colloquium on the Law of Outer Space*. IISL, Budapest, 1983.

Galloway, Jonathan F. "The Outer Space Treaty: 1967-2007". *Proceedings of the Fiftieth Colloquium on the Law of Outer Space*. IISL, Hyderabad, 2007.

Gorove, Katherine M. "Delimitation of Outer Space and the Aerospace Object – Where is the Law?" *Journal of Space Law*, Vol. 28, n. 11. Mississippi, USA, 2000.

Gorove, Stephen. "On the Threshold of Space: Toward a Cosmic Law – Problems of the Upward Extent of Sovereignty". *Proceedings of the First Colloquium on the Law of Outer Space*. IISL, The Hague, 1957.

Haanappel, Peter P. C. "Airspace, Outer Space and Mesospace". *Proceedings of the Nineteenth Colloquium on the Law of Outer Space*. IISL, New York, 1977.

Harris, Alexandra; Harris, Ray. "The Need for Air Space and Outer Space Demarcation". *Space Policy*. Vol. 22, n. 1. Dayton, USA, February 2006.

Heinrich, Welf. "Problems in Establishing a Legal Boundary between Air Space and Outer Space". *Proceedings of the First Colloquium on the Law of Outer Space*. IISL, The Hague, 1957.

Heraud, Bernard. "The Problem of the Delimitation of the Outer Space". *Proceedings of the Twenty-Second Colloquium on the Law of Outer Space*. IISL, Munich, 1979.

Hingorani, R. C. "La Souveraineté sur l'Espace Extra-Atmosphérique". *Revue Générale de l'Air*, Vol. 20. Paris, France, 1957.

Hobe, Stephen. "Outer Space as the Province of Mankind – An Assessment of 40 Years of Development". *Proceedings of the Fiftieth Colloquium on the Law of Outer Space*. IISL, Hyderabad, 2007.

Hogan, J. C. "Legal Terminology for the Upper Regions of the Atmosphere and Space Beyond the Atmosphere". *American Journal of International Law*, n. 51. New York, USA, 1957.

Hosenball, S. Neil; Hofgard, Jefferson S. "Delimitation of Air Space and Outer Space: Is a Boundary Needed Now". *University of Colorado Law Review*, Vol. 57. Boulder, USA, 1986.

Houmborg, Robert. "Etendues et limites du Droit Aérien". *Revue Générale de l'Air.* Paris, France, 1956.

Houmborg, Robert. "Droit Astronautique et Droit Aérien". *Revue Générale de l'Air.* Paris, France, 1958.

Hughes, William J. "Aerial Intrusions By Civil Airliner And The Use of Force". *Journal of Air Law and Commerce.* Dallas, USA, 1980.

Jacobini, H. B. "Effective Control as Related to the Extension of Sovereignty in Space". *Journal of Public Law*, n. 7, 1958.

Jastrow, Robert. "Definition of Air Space". *Proceedings of the First Colloquium on the Law of Outer Space*. IISL, The Hague, 1957.

Jayaraj, C. "Is There a Need for a Comprehensive Convention on Outer Space Law?" *Proceedings of the Fiftieth Colloquium on the Law of Outer Space*. IISL, Hyderabad, 2007.

Kopal, Vladimir. "Two Problems of Outer Space Control: the Delimitation of Outer Space, and the Legal Ground for Outer Space Flights". Proceedings of the Third Colloquium on the Law of Outer Space". *Proceedings of the Forth Colloquium on the Law of Outer Space*. IISL, Stockholm, 1960.

Kopal, Vladimir. "What is 'Outer Space' in Astronautics and Space Law". *Proceedings of the Tenth Colloquium on the Law of Outer Space*. IISL, Belgrade, 1967.

Kopal, Vladimir. "The question of defining of outer space". *Journal of Space Law*, Vol. 8. Mississippi, USA, 1980.

Kopal, Vladimir. "Issues Involved in Defining Outer Space, Space Object and Space Debris". *Proceedings of the Thirty-Forth Colloquium on the Law of Outer Space*. IISL, Montréal, 1991.

Kostantinov, Emil. "Some Aspects of the Spacial and Functional Delimitation between International Air and Space". *Proceedings of the Twenty-Sixth Colloquium on the Law of Outer Space*. IISL, Budapest, 1983.

Kroell, Joseph. "Éléments Créatures d'un Droit Astronautique". *Revue Générale de l'Air*, n. 16. Paris, France, 1953.

La Pradelle, Paul de. "Le Frontiére de l'Air". *Recueil de Cours*, II. The Hague, the Netherlands, 1954.

Lachs, Manfred. "Freedoms of the Air – the Way to Outer Space". *In* Masson-Zwaan, Tanja e De Leon, Pablo Mendes. *Air and Space Law: De Lege Ferenda – Essays in Honour of Henri A. Wessenbergh*. Dordrecht, the Netherlands: Martinus Nijhoff, 1992.

Lafferranderie, Gabriel. "Towards an Integrated Aerospace Law? Reflexions on the Definition and Delimitation of Outer Space", *In*: Brunner, Christian; Ulrich, Silvia e Lauer, Andrea. *Funktionen des Rechts in der Pluralistischen Wissensgesellschaft*. Vienna, Austria: Böhlau, 2007.

Lee, Ricky J e Eylward, Felicity K. "Article II of the Outer Space Treaty and Human Presence on Celestial Bodies: Prohibition of State Sovereignty, Exclusive Property Rights, or Both?". *Proceedings of the Forty-Eighth Colloquium on the Law of Outer Space*. IISL, Fukuoka, 2005.

Leister, Valnora. "O Comitê para o Uso Pacífico do Espaço Exterior (COPUOS) da Organização das Nações Unidas", *In*: Araminta Mercadante e José Carlos de Magalhães (coord.). *Reflexões sobre os 60 Anos da ONU*. Ijuí, Brazil: Unijuí, 2005.

Malenovsky, Jiri. "'Right' of Passage into Outer Space". *Proceedings of the Thirty-Third Colloquium on the Law of Outer Space*. IISL, Dresden, 1990.

Mapelli, Enrique. "The Difficult Delimitation of Airspace and Outer Space", *In*: Górbiel, Andrzej e Palyga, Edward J. *International Space Law Miscellanea: Liber Amicorum Honouring Professor Dr Andrzej Górbiel in his 65th Anniversary*. Warsaw, Poland: Andrzej Frycz Modrzewski Foundation, 1995.

Mishra, S. e Pavlasek, M. "On the Lack of Physical Bases for Defining a Boundary between Air Space and Outer Space". *Annals Air and Space Law*, Vol. VII. Montréal, Canada, 1982.

Masson-Zwaan, Tanja. "The Influence of the Achievements and Failures of the Past on the Future of Outer Space Law", *In:* Masson-Zwaan, Tanja (ed.). *Space Law: Views of the Future*. Deventer, the Netherlands: Kluwer, 1988.

Masson-Zwaan, Tanja. "The Aerospace Plane", *In*: Masson-Zwaan, Tanja e De Leon, Pablo Mendes. *Air and Space Law: De Lege Ferenda – Essays in Honour of Henri A. Wessenbergh*. Dordrecht, the Netherlands: Martinus Nijhoff, 1992.

Masson-Zwaan, Tanja. "Regulation of Sub-orbital Space Tourism in Europe: A Role for EU/ EASA?" *Air and Space Law*, Vol. 35, n. 3. Leiden, the Netherlands, 2010.

Menter, Martin. "Relationship of Air and Space Law". *Proceedings of the Nineteenth Colloquium on the Law of Outer Space*. IISL, New York, 1977.

Merignac, A. "Le Domain Aérien Privé et Public et le Droit de l'Aviation en Temps de Paix et Guerre". *Revue Générale de Droit International Public*, Vol. XXI. Paris, France, 1914.

Meyer, Alex. "Die Staatshoheit im Luftraum und die Entwicklungen im Weltraum". *ZLRW*, Cologne, Germany, 1965.

Moon, Albert. "A Look at Airspace Sovereignty". *Journal of Air Law and Commerce*, n. 29. Dallas, USA, 1963.

Nasser, Salem Hikmat. "Consequências jurídicas da edificação de um muro no território palestino ocupado", *In*: Mercadante, Araminta and Magalhães, José Carlos de (orgs.). *Reflexões sobre os 60 anos da ONU*. Ijuí, Brazil: UNIJUÍ, 2005.

Neumann, Julia. "An Interpretation of the Outer Space Treaty after 40 years". *Proceedings of the Fiftieth Colloquium on the Law of Outer Space*. IISL, Hyderabad, 2007.

Oduntan, Gbenga. "The Never Ending Dispute: Legal Theories on the Spatial Demarcation Boundary Plane between Airspace and Outer Space". *Hertfordshire Law Journal*, 1(2). Hertfordshire, England, 2003.

Ortiz, Jairo A. Becerra. "A Survey of Colombia's New Outer Space Policy: Reforms in Colombian Law". *Acta Astronautica*, Vol. 63, n. 1-4. Washington D. C., USA, July-August 2008.

Papathanassiou, C. "Refléxions sur les Probléms Juridiques Posés par le Vol à Haute Altitude et le Vol Cosmique". *Revue Hellénique de Droit International*, Vol. 11, n. 3-4. Athens, Greece, 1958.

Pelton, Joseph N. "A New Integrated Global Regulatory Regime for Air and Space: The Needs for Safety Standards for the 'Protozone'". *2nd Manfred Lachs International Conference on Global Space Governance*. Montreal, Canada: McGill University, 2014.

Peng, Ming Min. "Le Vol à Grande Altitude et l'Article 1er de la Convention de Chicago". *Révue Franceise de Droit Aérien*. Paris, France: 1952, n°. 4.

Perek, L. "Delimitation of Air Space and Outer Space: Is It Necessary?", *In*: Centre for Research of Air and Space Law (ed). *Earth Oriented Space Activities and Their Legal Implications: Proceedings of the Symposium Held on October 15–16, 1981*. Montréal, Canada, 1981.

Potter, Pitman B. "International Law of Outer Space". *American Journal of International Law*. Washington D. C., USA, 1958.

Qizhi, He. "The Problem of Definition and Delimitation of Outer Space". *Journal of Space Law*. Vol. 10, n. 157. Mississippi, USA, 1982.

Qizhi, He. "Review of Definition Issues in Space Law in the Light of Developments of Space Activities". *Proceedings of the Twenty-Forth Colloquium on the Law of Outer Space*. IISL, Montréal, 1991.

Rangel, Vicente Marotta. "Codificação do Direito Espacial". *Parcerias Estratégicas*, n. 7. Brasília, Brazil, 1999.

Rinck, Gerd. "Recht in Weltraum". *ZLRW*, Cologne, Germany, 1960. p. 191/208.

Romano, Roberto. "Paz de Westfália", *In*: Demétrio Magnoli (org.). *História da Paz*. São Paulo, Brazil: Contexto, 2008.

Rosenfield, S. B. "Some Thoughts on the Distinction between Air Space and Outer Space". *Proceedings of the Twenty-Sixth Colloquium on the Law of Outer Space*. IISL, Budapest, 1983.

Rosenfield, S. B. "Where Air Space Ends and Outer Space Begins". *Journal of Space Law*, Vol. 7, n. 148. Mississippi, USA, 1979.

Sangalli, Laureline e Knudsen, David. "Rocket-Based Measurements of Ion Velocity, Neutral Wind, and Electric Field in the Collisional Transition Region of the Auroral Ionosphere". *Journal of Geophysical Research*, 2009; 114 (a4): A04306.

Schachter, Oscar. "Legal Aspects of Space Travel". *J. B. I. P. S.*, London, England, 1952.

Schoenborn, Walter. "La Nature Juridique du Territoire." *Recueil de Cours*, t. 30. The Hague, the Netherlands, 1929.

Sloap, George Paul. "Relationship of Air Law and Space Law – a view from the Space Shuttle, including its Internal and External Environments". *Proceedings of the Nineteenth Colloquium on the Law of Outer Space*. IISL, New York, 1977.

Smirnoff, M. "The Need for a New System of Norms for Space Law and the Danger of Conflict with the Terms of the Chicago Convention". *Proceedings of the First Colloquium on the Law of Outer Space*. IISL, The Hague, 1957.

Sterns, Patricia M. e Tennen, Leslie I. "Ethics and the Conquest of Space: From Peenemunde to Mars and Beyond". *Proceedings of the Fiftieth Colloquium on the Law of Outer Space*. IISL, Hyderabad, 2007.

Strughold, Hubert. "Definitions and Subdivisions of Space: Bioastronautical Aspect". *Proceedings of the First Colloquium on the Law of Outer Space*. IISL, The Hague, 1957.

Tang, C. S. "The Boundary Question in Space Law: a Balance Sheet". *Ottawa Law Review*, Vol. 6, n. 263. Ottawa, Canada, 1973.

Tennen, Leslie I. "Conflicts of Law and Delineation of Outer Space: an Interest Analysis Approach". *Proceedings of the Twenty-Second Colloquium on the Law of Outer Space*. IISL, Munich, 1979.

Terekchov, A. D. "Passage of Space Objects Through Foreign Airspace". *IISL*, Leiden, the Netherlands, 1989.

Tronchetti, Fabio. "The Non-Appropriation Principle Under Attack: Using Article II of the Outer Space Treaty in its Defence". *Proceedings of the Fiftieth Colloquium on the Law of Outer Space*. IISL, Hyderabad, 2007.

Valladão, Haroldo. "The Law of Interplanetary Space". *Proceedings of the Second Colloquium on the Law of Outer Space*. IISL, London, 1958.

Vázquez, Modesto Seara. "The Functional Regulation of the Extra-Atmospheric Space". *Proceedings of the Second Colloquium on the Law of Outer Space*. IISL, London, 1958.

Van Der Heijden, Marc A. J. M. "Ceiling of Sovereignty". *Thesauros Acroasium*. Institution of International Public Law and International Relations of Thessaloniki, Vol. XIV. Thessaloniki, Greece., 1985.

Vassilevskaya, E. G. "The Importance of the Delimitation for Air and Space Law". *Proceedings of the Twenty-Sixth Colloquium on the Law of Outer Space*. IISL, Budapest, 1983.

Vassilevskaya, E. G. "Delimitation of Air and Space – Major Issue for Ensuring Peace and Security". *Proceedings of the Twenty-Eighth Colloquium on the Law of Outer Space*. IISL, Stockholm, 1985.

Vereschaguin, A. "Derecho Aéreo Internacional", *In*: Tunkin, G. (coord.). *Curso de Derecho Internacional: Manual*. Moscow, Russia: Progresso, 1979.

Vereshchetin, Vladen. "New Steps in International Space Law", *In*: Jasentuliyana, Nandasiri. *Perspectives in International Law*. The Hague, the Netherlands: Kluwer, 1995.

Von der Dunk, Frans G. "The Delimitation of Outer Space Revisited – The Role of National Space Laws in the Delimitation Issue". *Proceedings of the Forty-First Colloquium on the Law of Outer Space*. IISL, Melbourne, 1998.

Von der Dunk, Frans G. "Launching from 'Down Under': the New Australian Space Activities Act of 1998". *Proceedings of the Forty-Third Colloquium on the Law of Outer Space*. IISL, Rio de Janeiro, 2000.

Von der Dunk, Frans G. "The Sky is the Limit – but Where does it End?". *Proceedings of the Forty-Eighth Colloquium on the Law of Outer Space*. IISL, Fukuoka, 2005.

Voûte, Caesar. "Boundaries in Space", *In*: Bhupendra Jasani (ed.). *Peaceful and Non-Peaceful Uses of Space*. New York, USA: Taylor & Francis, 1991.

Williams, Maureen. "The Problem of Demarcation is Back in the Limelight". *Proceedings of the Twenty-Second Colloquium on the Law of Outer Space*. IISL, Munich, 1979.

Wolczek, Olgierd. "Remarks". *Proceedings of the First Colloquium on the Law of Outer Space*. IISL, The Hague, 1957.

Zadorozhny, G. P. "Iskustvennye Sputniki i Mezhdunarodnoye Pravo". *Sovietskaya Rossiya*. Moscow, Russia, October 1957.

MIX
Papier aus verantwortungsvollen Quellen
Paper from responsible sources
FSC® C105338

FSC
www.fsc.org

If you have any concerns about our products,
you can contact us on
ProductSafety@springernature.com

In case Publisher is established outside the EU,
the EU authorized representative is:
Springer Nature Customer Service Center GmbH
Europaplatz 3, 69115 Heidelberg, Germany

Printed by Libri Plureos GmbH
in Hamburg, Germany